FLORA OF TROPICAL EAST AFRICA

DRACAENACEAE

GEOFFREY MWACHALA & PAUL MBUGUA*

Trees, shrubs or herbs. Leaves alternate, rosulate, distichous or spirally arranged, sometimes in pseudo-whorls, sessile, the base often (semi-)amplexicaul, succulent or not; primary venation parallel. Inflorescence usually a terminal panicle, occasionally appearing racemose by reduction of primary branches; pedicel of a peg-like basal part crowned by a distinct joint, and an obconical receptacle, persistent in fruit. Flowers in irregular clusters, 1–several per cluster, subsessile, with a bract subtending each flower, fragrant, nocturnal; perianth tubular, with 6 spreading or recurving lobes. Stamens 6, inserted in the throat, with inflated or filiform filaments; anthers dorsifixed, versatile, opening by lateral slits. Ovary superior, free, globose, 3-celled; ovules erect, solitary in each of the three chambers; style filiform; stigma capitate. Fruit a berry, 1–3-seeded. Seeds rounded, bony, white.

Leaves leathery, not fleshy; trees or shrubs, the stem usually several metres high	1. **Dracaena**
Leaves fleshy; usually herbs without visible stems (but some species with stem to 2.5 m high)	2. **Sansevieria**

1. DRACAENA

L., Syst. Nat. ed. 12, II: 246 (1767) & Mant. Pl. I: 63 (1767); Bos in Kubitzki et al., Fam. Gen. Vasc. Pl. 3: 240 (1998)

Pleomele Salisb., Prodr. 245 (1796)

Trees, shrubs, sometimes scandent, or herbs; roots often orange. Leaves alternate, distichous or spirally arranged, sometimes in pseudo-whorls, sessile, the base often (semi-)amplexicaul; primary venation parallel. Inflorescence usually a terminal panicle, occasionally appearing racemose by reduction of primary branches; pedicel of a peg-like basal part crowned by a distinct joint, and an obconical receptacle, persistent in fruit. Flowers fragrant, nocturnal; tubular, with 6 recurving lobes. Stamens inserted in the throat, with inflated filaments; anthers versatile, opening by lateral slits. Style filiform. Fruit a berry, 1–3-seeded. Seeds rounded, bony, white.

At least 60 species, mainly in Africa and SE Asia.

* GM: *Dracaena*. Address: East African Herbarium, P.O.Box 45166, Nairobi 00100, Kenya
 PM: *Sansevieria*. Address: Kenyatta University, Department of Plant and Microbial Sciences, P.O. Box 43844 00100 GPO, Nairobi, Kenya
With thanks to Mark Coode for translations into Latin, and Adelaide Stork for several comments

CULTIVATED TAXA

Dracaena angustifolia Roxb.; U.O.P.Z.: 236 (1949). A weak, much-branched shrub with sessile linear leaves in dense rosettes, 15–25(–49) cm long, 1–3.5 cm wide. Inflorescence terminal, paniculate, usually with only one order of branching; pedicels 0.5–1 cm long, articulated above the middle; perianth greenish white, 1.8–2 cm long.

From south-east Asia, reported in cultivation in Pemba in 1949 (U.O.P.Z.) I have not seen any herbarium material from our area.

Dracaena bicolor Hook. (Syn.: *Dracaena cylindrica* Hook.f.) U.O.P.Z.: 238 (1949). Shrub with erect, unbranched stem 3–4 m high; leaves oblanceolate, 15–25(–50) cm long, 5–10 cm wide, pseudopetiolate, apex acute with a tiny mucro. Inflorescence terminal, with dense clusters of flowers subtended by purplish 2–5 cm long bracts, pedicels 2–3 mm long, jointed at the middle; perianth white, purple-tinged, 25–28 mm long. Fruits orange, drying black.

Dracaena draco L. Profusely branched tree to 20 m tall. Leaves to 1.5 m long, 3–6 cm wide, sessile, ensiform. Inflorescence terminal, paniculate. Inflorescence branches 18–36 cm long; pedicels 7–12 mm long, articulated at the middle. Perianth green, to 1 cm long. Fruit orange, 1-seeded.

Cultivated at Amani (29 Oct. 1928, *Greenway* 958!), and at the Village Market shopping centre in Nairobi. Exotic from the Canary Islands, grown as an ornamental on account of its unusual palm-like appearance.

1. Leaves pseudo-petiolate; herbs and shrubs 2
   Leaves sessile, non-pseudo-petiolate; habit various 3
2. Erect shrubs with distichous leaves; flower clusters on
   distinct knobs . . . . 1. *D. aubryana*
   Sprawling, sarmentose or scrambling shrubs; flowers
   single, not on distinct knobs . . . . 2. *D. laxissima*
3. Leaves whorled; the leafless nodes in between the leaf
   whorls bearing prophylls . . . . 3. *D. camerooniana*
   Leaves clustered at end of branches or spaced out along
   the branches, not whorled; leafless nodes usually below
   the leafy ones, not bearing prophylls 4
4. Leaves linear, ensiform, widest at the base; flowers single
   or in clusters of up to 3 . . . . 4. *D. ellenbeckiana*
   Leaves linear-lanceolate, widest above the bottom quarter
   of their length; inflorescence various 5
5. Leaves linear-lanceolate (sword-shaped), at least 5 cm
   wide, becoming pendulous with age; perianth lobes
   shorter than tube; inflorescence paniculate . . . . 5. *D. aletriformis*
   Leaves narrowly lanceolate to oblanceolate or narrowly
   elliptic, less than 5 cm wide; perianth lobes longer than
   tube; inflorescence various 6
6. Leaves narrowly lanceolate, 12–35 cm long, 1.5–3 cm
   wide; inflorescence pendent; flower clusters of no more
   than 4 . . . . 6. *D. afromontana*
   Leaves narrowly oblanceolate or narrowly elliptic;
   inflorescence erect; flower clusters of 5 or more 7
7. Leaves narrowly oblanceolate; perianth 11–25 mm long,
   yellow-green, mauve or purple; fruit orange or purple
   when ripe 8
   Leaves narrowly elliptic; perianth 30–40 mm long, yellow-
   white on the outside; fruit red, to 3 cm in diameter . . . 7. *D. mannii*

8. Leaves clustered at ends of branches; inflorescence 0.3–1 m
long, up to 1 m wide; flowers white, cream or greenish
yellow; ripe fruits purple . . . .   8. *D. steudneri*
Leaves spaced out along the branches; inflorescence
0.3–1.6 m long, to 0.6 m wide; flowers mauve to purple,
at least on the outside, the inside usually white; ripe
fruits bright orange . . . .   9. *D. fragrans*

1. **Dracaena aubryana** *E.Morren*, Belg. Hort. 10: 348 (1860) & in Bull. Fed. Soc.
1860: 302 (1861); Bos, Dracaena in W Afr.: 29, t. 4, photo. 10 (1984). Type: cultivated
in Hort. Paris, 1859, *Wentzel* in herb. *Martius* (BR, lecto., chosen by Bos, 1984)

Shrub 0.4–2.5 m high, usually unbranched; stem erect, often twisted spirally.
Leaves distichous, often asymmetric, ovate to narrowly ovate, 10–40(–60) cm long,
(1.5–)4–10(–15) cm wide, base rounded or cuneate, apex acute to cuspidate; petiole
5–30(–100) cm long. Inflorescence erect, spicate or paniculate, 9–70 cm long,
unbranched or with a few branches near the base; flowers in groups of 1–3(–7) on
distinct knobs with a single triangular bract to 10 mm long; pedicel 0–2 mm long,
articulated below the middle. Flowers white or greenish-white, each lobe often with
purple-red central band, (10–)15–30(–55) mm long, tube 5–10(–30) mm long, lobes
10–20 mm long, to 2.5 mm wide. Fruits shiny bright orange, deeply 1–3-lobed, lobes
ovoid, 8–18 mm long, 4–9 mm in diameter; seeds dirty white, ovoid, 6–14 mm wide,
5–7 mm long.

UGANDA. Toro District: Bwamba, Feb. 1939, *A.S. Thomas* 2784! & May 1989, *Katende* 3690!
DISTR. U 2; West and Central Africa from Sierra Leone to Congo-Kinshasa and Angola
HAB. Moist forest; may be locally common; 600–850 m
USES. None recorded
CONSERVATION NOTES. Least concern
SYN. *Dracaena thalioides* Regel, Gartenflora 20: 147 (1871); Baker in F.T.A. 7: 445 (1898) pro
     parte; F.W.T.A. ed. 2, 3 (1): 156 (1968). Type: Belg. Hort. 10: t. 24 (1860), iconotype
   D. humilis Baker in J. Bot. 12: 166 (1874) & in J.L.S. 14: 534 (1875) & in F.T.A. 7: 444
     (1898); F.W.T.A. ed. 2, 3 (1): 156 (1968). Type: Sierra Leone, Bagroo R., *Mann* 898 (K!,
     holo., A, P, iso.)
   D. kindtiana De Wild. in Ann. Mus. Congo scr. 5, 2: 119, t. 65, 66 (1907). Type: t. 65 in Ann.
     Mus. Congo ser. 5, 2 (1907), iconotype
   Pleomele thalioides (Regel) N.E.Br. in K.B. 1914: 279 (1914). Type as for *D. thalioides* Regel

2. **Dracaena laxissima** *Engl.* in E.J. 15: 478 (1892); Baker in F.T.A. 7: 446 (1898);
T.T.C.L.: 21 (1949); F.W.T.A. ed. 2, 3 (1): 157 (1968); Hamilton, Uganda forest
trees: 78 (1981); Troupin, Fl. Pl. Lign. Rwanda: 68 (1982); Bos, Dracaena in W Afr.:
79, t. 14 (1984); K.T.S.L.: 640 (1994). Type: Congo-Kinshasa, Mukenge, *Pogge* 1462
(B, holo.)

Shrub, 1–6 m high, often sprawling, sarmentose or scrambling, the stem
prominently lenticillate. Leaves elliptic or less often ovate or obovate, (4–)8–20 cm
long, (1–)2.5–7 cm wide, base cuneate to rounded into an amplexicaul sheath, apex
acuminate; petiole 5–15 mm long. Inflorescence horizontal to pendent, paniculate,
(5–)15–50 cm long, flowers occurring singly, each with a sheathing bract 2–3 mm
long; pedicel 2–17 mm long, articulated above the middle. Flowers white, greenish
white or greenish pink, 12–20 mm long, tube to 3 mm, lobes 9–17 mm long, 2.5 mm
wide, with a single median rib. Fruits orange-red, globose, lobed when several-seeded,
6–12 mm in diameter; seeds globose or flattened, 4–7 mm in diameter.

UGANDA. Ankole District: Kalinzu Forest, July 1938, *Eggeling* 3783!; Kigezi District: Murole Hill,
     Apr. 1948, *Purseglove* 2692!; Masaka District: Bukoto, Bugonzi, July 1971, *Katende* 1165!

KENYA. Northern Frontier District: Marsabit, Jan. 1961, *Polhill* 345!; Nairobi, Karura Forest, May 1949, *Bogdan* 2445!; Teita District: Sagala Hill, June 1985, *Taita Hills Exped.* 1076!
TANZANIA. Lushoto District: Amani road to Derema, June 1970, *Kabuye* 193!; Morogoro District: N Nguru Mts, Mt Kanga, Dec. 1987, *Lovett & Thomas* 2680!; Mufindi District: Rufuna Forest Reserve, Nov. 1982, *Leliyo* 346!; Pemba: Ngezi Forest, 1929, *Taylor* 96/2!
DISTR. U 2–4; **K** 1, 4–7; **T** 2–3, 6–8; widespread in tropical Africa
HAB. Moist forest, in dry forest usually near water, riverine forest; may be locally common; 1–2250 m
USES. Root decoction as minor medicine
CONSERVATION NOTES. Least concern

SYN. *Dracaena elegans* Hua in Contr. Fl. Congo fr., Lil: 13 (1897); Baker in F.T.A. 7: 446 (1898). Type: Gabon, Sanga, *Leroy* s.n. (P, holo., K!, iso., drawing)
    Pleomele elegans (Hua) N.E.Br. in K.B. 1914: 278 (1914)

3. **Dracaena camerooniana** *Baker* in J.B. 12: 166 (1874) & J.L.S. 14: 538 (1875) & in F.T.A. 7: 442 (1898); F.F.N.R.: 16 (1962); F.W.T.A. ed. 2, 3 (1): 157 (1968); Bos, Dracaena in W Afr.: 45, t. 7, 8 (1984). Type: Cameroon, Mt Cameroon, *Mann* 1204 (K!, holo.; A, P, iso.)

Shrub 0.3–3.5(–8) m high; stems branched, producing cane-like shoots, leafy or clad in caducous white or pale green prophylls with distant pseudowhorls of leaves; prophylls triangular, to 10 cm long, sheathing the stem; branches at right angles to the stem, piercing the sheathing base of the supporting leaf. Leaves obovate, 5–33 cm long, 1–8.5 cm wide, base cuneate to attenuate into the 1–4 cm long petiole, the very base amplexicaul, apex acuminate. Inflorescence pendent, spicate, 5–50 cm long; flowers in groups of 2–20; pedicels to 5 mm long, articulated above the middle. Flowers white or with some purple on the lobes, 1.9–3 cm long, tube as long as or twice or more as long as the lobes, tube 15–20 mm long, lobes 4–10 mm long, ± 2 mm wide. Fruits orange or red, globose or depressed globose, lobed when more than 1-seeded, 7–21 mm in diameter; seeds straw-coloured, hemispherical when single, flattened when several, 4–11 mm long, 5–14 mm wide and 2–9 mm thick.

TANZANIA. Ufipa District: Kitungulu, *Munzner* in exped. *Fromm* 245!
DISTR. **T** 4; W Africa from Guinea to Central African Republic, Democratic Republic of Congo and S to Angola and Zambia
HAB. Riverine forest; ± 1500 m
USES. None recorded
CONSERVATION NOTES. Least concern

SYN. *Dracaena capitulifera* De Wild. & Th. Durand in Ann. Mus. Congo 1, 1: 59 (1899). Type: Congo-Kinshasa, Bolobo, *Dewèvre* 701 (BR, holo.)
    D. gentilii De Wild., Et. Fl. Bas- et Moyen Congo 1, 3: 228 (1906). Type: Congo-Kinshasa, Luebo-Luluabourg, *Gentil* 54 (BR, holo.)
    D. uleensis De Wild., Et. Fl. Bas- et Moyen Congo 2: 20, t. 8, 9 (1907). Type: Congo-Kinshasa, Suronga, *Seret* 397 (BR, holo.)
    D. frommii Engl. & K.Krause in E.J. 45: 151 (1910). Type: Tanzania, Kitungulu, *Munzner* in exped. *Fromm* 245 (B, holo.)
    D. dundusanensis De Wild. in B.J.B.B. 5: 5 (1915). Type: Congo-Kinshasa, Dundusana, *De Giorgi* 1069 (BR, lecto., chosen by Bos)

4. **Dracaena ellenbeckiana** *Engl.* in E.J. 32: 95 (1902); K.T.S.L.: 639, fig. (1994); Bos in Fl. Eth. & Eritr. 6: 77 (1997). Type: Ethiopia, Luku–Sheik Hussein, *Ellenbeck* 1232 (B, holo.; EA!, iso., K!, photo.)

Shrub or tree 2–8 m high; stems erect, several from a common base, less often solitary, little-branched, up to 8 cm in diameter, longitudinally fissured; branches pale grey with a reticulate pattern of leaf scars. Leaves clustered at branch ends,

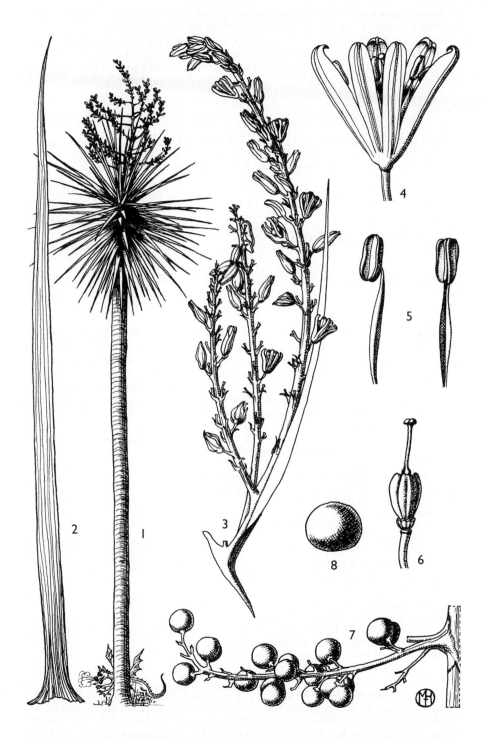

FIG. 1. *DRACAENA ELLENBECKIANA* — **1**, habit, × ⅛; **2**, leaf, × ½; **3**, inflorescence, × 1; **4**, flower, × 4; **5**, stamens, × 8 ; **6**, pistil, × 4; **7**, fruit, × ⅔; **8**, seed, × 2. Drawn by Olive Milne-Redhead.

narrowly lanceolate, sessile, 22–65 cm long, 1–9 cm wide, base clasping, apex acuminate. Inflorescence erect, paniculate, 20–80 cm long with 2–3 branches at each node; flowers in groups of (1–)2–7, pedicels 2–5 mm long, articulated above the middle. Flowers white becoming yellow-green, to 10 mm long, tube 2 mm long, lobes 8 mm long, translucent, 1-ribbed, 1–1.5 mm wide. Fruits orange to scarlet, globose or lobed when several-seeded, 8–14 mm in diameter; seeds brown, globose, ± 6 mm in diameter. Fig. 1, p. 5.

UGANDA. Karamoja District: Kamion, Nov. 1939. *A.S. Thomas* 3254!; & NE Karamoja, June 1945, *Dale* 432!
KENYA. Turkana District: Kachagalau, E slope of Kalapata Mts, Jan. 1959, *Langdale Brown* 95!; Laikipia District: 30 km N of Rumuruti, Kisima Farm, Apr. 1975, *Hepper & Field* 5078!; Kiambu District: Kikuyu Escarpment just below lower Lari Forest guard post, Dec. 1966, *Perdue & Kibuwa* 8259!
DISTR. **U** 1; **K** 1–4, 6; Sudan, Ethiopia
HAB. Semi-evergreen bushland or open dry forest on rocky slopes; may be locally common; 1050–2100 m
USES. Stems used for arrow quivers
CONSERVATION NOTES. Least concern

5. **Dracaena aletriformis** (*Haw.*) Bos in F.S.A. 5, 3: 3 (1992); K.T.S.L: 639 (1994). Type: South Africa, Cape, Uitenhage, van Stadens R., *Drège* 4494a (K!, neo., G, MO, P, isoneo.), chosen by Bos

Small tree 3–5 m high; stem erect, branched. Leaves congested towards branch ends, narrowly to broadly strap-shaped, the oldest ones pendent, sessile, 25–100 cm long, 2.5–11 cm wide, with a scarious edge ± 1 mm wide; base sheathing, apex acute, tapering gradually into a filiform tip. Inflorescence erect, paniculate, 0.2–1.5 m long; flowers in groups of 1–4, more in terminal clusters; pedicel 5–10 mm long, articulated above the middle. Flowers greenish white, sometimes flushed with pink, 19–35(–42) mm long; tube 7–12 mm long, lobes 12–20 mm long, 1.5–2 mm wide, lobes ± 1.5 × as long as the tube. Fruit orange or red, usually globose and single-seeded, sometimes lobed and 2–3-seeded, 7–19 mm in diameter; seed dirty white, ovoid, 5–10 mm long, 8–12 mm wide.

KENYA. Kilifi District: Rabai, Apr. 1963, *Verdcourt* 3606! & Kaya Kambe, July 1987, *Robertson & Luke* 4781!; Kwale District: Mrima Hill, 2 Feb. 1989, *Schmidt* 1712!
TANZANIA. Zanzibar, 1852, *Boivin* s.n.!
DISTR. **K** 7; **Z**; Mozambique, South Africa
HAB. Evergreen or semi-deciduous forest; 50–300 m
USES. None recorded
CONSERVATION NOTES. Rather uncommon in our area; its habitat, bushland on coral rag, is under threat from human settlements.

SYN. *Yucca aletriformis* Haw. in Phil. Mag. Ann. 1832: 415 (1831)
     Cordyline rumphii Hook. in Bot. Mag. 3rd ser., 3: t. 4279 (1847) pro parte, quoad descr. excl. syn., *nom. illeg.* Type: t. 4279 (lecto.), chosen by Bos
     Dracaena hookeriana K.Koch in Wochenschr. Gartn. Pflanzenk. 4: 394 (1861); Baker in J.L.S., Bot. 14: 527 (1875) & Fl. Cap. 6: 275 (1896); Coates Palgrave, Trees S Afr.: 86 (1977), *nom. illeg.* Type as for *Cordyline rumphii*
     Pleomele hookeriana (K.Koch) N.E.Br. in K.B. 1914, 8: 278 (1914)

6. **Dracaena afromontana** Mildbr., Z.A.E.: 62, t. V (1910); T.T.C.L.: 20 (1949); Hamilton, Uganda forest trees: 78 (1981); Troupin, Fl. Pl. Lign. Rwanda: 68 (1982); K.T.S.L.: 639 (1994); Bos in Fl. Eth. & Eritr. 6: 76, t. 185/1–4 (1997). Types: Rwanda, Rugege Forest, *Mildbraed* 1033; Ninagongo, *Mildbraed* 1360; Ruwenzori, *Mildbraed* 2525 (all B†, syn.); neotype: Z.A.E.: 63, t. V (1914), chosen by Bos, 1997

Shrub or tree, 2–12 m high, sometimes straggling; stem to 25 cm in diameter, branches few, arching, with a distinct pattern of horseshoe-shaped leaf-scars. Leaves narrowly lanceolate, sessile, 12–35 cm long, 1.5–3 cm wide, base hardly constricted and clasping, apex acute. Inflorescence pendulous, paniculate, 20–60 cm long; flowers in groups of 1–3; pedicels 4–12 mm long, articulated above the middle. Flowers white, pale green or with a purple tinge outside, somewhat translucent, 1-ribbed, 15 mm long; tube 1 mm long, lobes 14 mm long, 2–3 mm wide, spreading at anthesis. Fruits orange, globose or lobed when several-seeded, 12–20 mm in diameter; seeds dirty white, 6–9 mm in diameter.

UGANDA. Karamoja District: Saosa catchment near Mt Moroto, Jan. 1959, *Kerfoot* 715!; Toro District: Ruwenzori foothills, Sep. 1936, *Mukibi* in *A.S. Thomas* 2627!; Elgon, Oct. 1923, *Snowden* 804!
KENYA. Northern Frontier District: Mt Kulal, Narangani, June 1960, *Oteke* 97!; Trans Nzoia District: Cherangani Mts, Kabolet, Aug. 1963, *Tweedie* 2709!; Mt Kenya, 3 km below Castle Forest Station, Jan. 1967, *Perdue & Kibuwa* 8361!
TANZANIA. Kilimanjaro, above Kilimanjaro Timbers, May 1994, *Grimshaw* 94/519!; Ufipa District: Mbizi Forest Reserve, Nov. 1987, *Ruffo & Kisena* 2871!; Njombe District: Ndumbi Forest, Mtorwi, Nov. 1986, *Goldblatt & Lovett* 8236!
DISTR. **U** 1–3; **K** 1, 3–4, 6–7; **T** 2–4, 6–7; Congo-Kinshasa, Rwanda, Burundi, Ethiopia, Malawi
HAB. Rain-forest, moist forest, or bamboo forest, often forming dense stands in the understorey; 1600–2700 m
USES. None recorded
CONSERVATION NOTES. Least concern

7. **Dracaena mannii** *Baker* in J.B. 12: 164 (1874) & in J.L.S. 14: 526 (1875) & in F.T.A. 7: 438 (1898); F.W.T.A. ed. 2, 3 (1): 156 (1968); Bos, Dracaena in W Afr.: 82, t. 15 & 16, 12 (1984) & in F.S.A. 5, 3: 3, t. 9 (1992). Type: Nigeria, Old Calabar, *Mann* 2329 (K!, lecto., chosen by Bos, 1984; A, B, P, WAG, iso.)

Shrub or tree (2.4–)4.5–30 m high, evergreen, single-stemmed or often multistemmed or much branched from near ground level; stem to 2 m in diameter. Leaves clustered near branch ends, narrowly elliptic to narrowly lanceolate or obovate, sessile, (4–)8–40(–50) cm long, (0.3–)0.8–3.8(–4.5) cm wide, base cuneate but the very base half-amplexicaul, apex acute with a mucro, 'midrib' prominent beneath. Inflorescence erect, racemose to paniculate, continous to stem and branches; flowers in groups of 2–6, rarely solitary; pedicel 1–4 mm long or up to 9 mm and orange in fruit, articulated at the top. Flowers creamy white or pure white, yellow-green outside, the tips of lobes greenish to purple, 12–50 mm long, tube 3–25 mm long, lobes 8–24 mm long, 2–3 mm wide. Fruits orange to scarlet, globose or lobed when 2–3-seeded, 10–32 mm in diameter; seeds brown, globose, 10–20 mm in diameter.

KENYA. Kilifi District: Kaya Chonyi, Sep. 1990, *Robertson* 6391!; Kwale District: Shimba Hills, Pengo Hill, May 1968, *Magogo & Glover* 1061! & S end of Shimba Hills Reserve, Apr. 1977, *Gillett* 21075!
TANZANIA. Lushoto District: E Usambara Mts, Kiwanda, Oct. 1936, *Greenway* 4698!; Kilosa District: Vigude Forest, Kidodi, Nov. 1952, *Semsei* 1006!; Uzaramo District: Mnguvia R. 16 km SE of Dar es Salaam, Sep. 1977, *Wingfield* 4120!
DISTR. **K** 7; **T** 3, 4, 6–7; **Z**; widespread in tropical Africa
HAB. Moist forest, often near streams, or riverine forest, also in forest margins and secondary bushland; may be locally common; 0–1400(–1600) m
USES. Branches used for toothbrushes
CONSERVATION NOTES. Least concern
NOTE. Bos (1984) cites *Mann* 2339 as the lectotype, this is clearly a typographical error, as the specimen he annotated as lectotype at K is *Mann* 2329.

SYN. *Dracaena usambarensis* Engl. in Abh. Preuss. Akad. Wiss. 1894: 30, nomen (1894) & P.O.A. C: 144 (1895); T.T.C.L.: 21 (1949); K.T.S.L.: 639 (1994). Type: Tanzania, Lushoto District: Mto wa Simbili between Magila and Sigi R., *Volkens* 65 (B†, syn., K!, photo.) & Schira, *Volkens* 1938 (B†, syn., BR, K!, photo)

D. pseudoreflexa Mildbr. in Z.A.E.: 63, t. 5/g–k. (1914). Type: Congo-Kinshasa, between Beni
and Irumu, *Mildbraed* 2813 (B, holo.; K, photo!)
Pleomele mannii (Baker) N.E.Br. in K.B. 1914: 278 (1914)
P. usambarensis (Engl.) N.E.Br. in K.B. 1914: 279 (1914)

8. **Dracaena steudneri** Engl., P.O.A. C: 143 (1895); Baker in F.T.A. 7: 441 (1898);
T.T.C.L.: 21 (1949); Hamilton, Uganda forest trees: 76 (1981); Troupin, Fl. Pl. Lign.
Rwanda: 68 (1982); K.T.S.L.: 640 (1994); Bos in Fl. Eth. & Eritr. 6: 79, t. 187/5 (1997).
Type: Ethiopia, Gondar, Dschibba, *Steudner* 477 (B†, holo.; S, iso., BR, K, photo!)

Tree (3–)6–15(–25) m high; stem single, to 45 cm in diameter; crown branched.
Leaves clustered at branch ends, leathery, narrowly lanceolate, sessile, 40–130 cm
long, 4–16 cm wide, base narrowed and clasping, apex attenuate towards acute tip.
Inflorescence erect becoming pendulous in fruit, paniculate, 0.3–1(–2) m long and
up to 1.5 m wide; flowers in almost globose groups of up to 100 flowers; pedicels
2.5 mm long, articulated above the middle. Flowers white, cream or greenish white,
11–18 mm long, tube 6–10 mm long, lobes 5–8 mm long and 0.5–1 mm wide,
translucent, 1-ribbed. Fruits bronze to dark purple to blackish red, 12–30 mm in
diameter; pulp orange; seeds white, globose, 10 mm in diameter.

UGANDA. Karamoja District: Mt Lonyili, Kidepo National Park, 16 May 1972, *Synnott* 1022;
Kigezi District: Kachwekano Farm, Sep. 1949, *Purseglove* 3119!; Mengo District: Kampala, Sep.
1930, *Snowden* 1821!
KENYA. Trans Nzoia District: Elgon, 15 Mar. 1933, *Jack* 427; Kiambu District: Karura Forest, 14
Dec. 1966, *Perdue & Kibuwa* 8239!; Teita District: Mt Kasigau, path from Rukanga up the
mountain, 5 Apr. 1969, *Faden et al.* 459!
TANZANIA. Lushoto District: Lushoto, Oct. 1963, *Semsei* 3644!; Ulanga District: Ujiji near
Mahenge, Oct. 1960, *Haerdi* 595/0!; Songea District: Luwira-Kitega Forest Reserve, Oct. 1956,
Semsei 2536!
DISTR. **U** 1, 2, 4; **K** 3, 4, 7; **T** 3, 6–8; Congo-Kinshasa, Ethiopia and S to Mozambique and
Zimbabwe
HAB. Moist or dry forest or forest margins, sometimes planted as a boundary in fields;
850–2300 m
USES. Leaves sometimes used as wrapping material; leaves and root in minor medicine
CONSERVATION NOTES. Least concern

SYN. *Dracaena papahu* Engl., P.O.A. C: 143 (1895). Type: Tanzania, Lushoto District: Lutindi,
Usambara, *Holst* 3260 (K!, iso.)
Pleomele papahu (Engl.) N.E.Br. in K.B. 1914: 278 (1914)

9. **Dracaena fragrans** (*L.*) *Ker-Gawl.* in Bot. Mag.: t. 1081 (1808); Baker in J.B. 12:
165 (1874) & in J.L.S. 14: 529 (1875) & in F.T.A. 7: 440 (1898); T.T.C.L.: 21 (1949);
U.O.P.Z.: 236 (1949); F.W.T.A. ed. 2, 3 (1): 157 (1968); Hamilton, Uganda forest
trees: 78 (1981); Bos, Dracaena in W Afr.: 69, t. 13, photo 12 (1984); K.T.S.L.: 639
(1994); Bos in Fl. Eth. & Eritr. 6: 77 (1997). Type: Commelin, Hort. Med. Amst. 2: t.
4 f. 2 (1701)

Shrub or tree, 1.5–15 m high; stem single, up to 30 cm in diameter, or in forest
often with a mass of horizontal stems near the ground from which vertical stems
arise. Leaves strap-shaped to oblanceolate, sessile, (7–)20–150 cm long, 2–12 cm
wide, base attenuate, widening for 2–5 cm and sheathing the stem, apex acute with
subulate mucro. Inflorescence erect, rarely pendent, paniculate, 15–160 cm long;
flowers in multiflowered glomerules; pedicels 2–6 mm long, articulated below the
middle. Buds purple or pink, flowers with tube white, the lobes white with a fine red
to purple line down the centre, darker outside, (15–)17–22(–25) mm long, tube
5–11 mm long, lobes 7–12 mm long, up to 3 mm wide. Fruits bright orange,
depressed globose, 11–19 mm in diameter, lobed when several-seeded; seeds white
turning brown, globose to bean-shaped, 4–14 mm in diameter.

UGANDA. Toro District: Kibale Forest, Sebutole, Feb. 1953, *Osmaston* 2812!; Ankole District: Igara, Mar. 1939, *Purseglove* 637!; Mengo District: 14 km on Masaka road, Oct. 1937, *Chandler* 1942!

KENYA. North Kavirondo District: near Kakamega Forest Station, Dec. 1967, *Perdue & Kibuwa* 9400!; Kericho District: Sotik, Kibajet Estate, Dec. 1947, *Bally* 5720!; Teita District: Kasigau, Rukanga route, 16 Nov. 1994, *Luke & Luke* 4166!

TANZANIA. Bukoba District: Kagera, Minziro, Feb. 1995, *Congdon* 413!; Tanga District: Potwe Forest, Dec. 1960, *Semsei* 3187!; Morogoro District: Kanga Mts, 90 km N of Morogoro, Dec. 1987, *Mwasumbi & Munyenyembe* 13875!

DISTR. **U** 2–4; **K** 1–7; **T** 1–8; throughout tropical Africa

HAB. Moist forest, often near streams; may be locally common and may form thickets; 600–2250 m

USES. Used for hedges/living fences

CONSERVATION NOTES. Least concern

SYN. *Aletris fragrans* L., Sp. Pl. ed. 2: 456 (1762)
    Pleomele fragrans (L.) Salisb., Prodr.: 245 (1796); N.E. Br. in K.B. 1914: 276, 278 (1914)
    Sansevieria fragrans (L.) Jacq., Fragm. Bot: 5, t. 2 f.6 (1800)
    Cordyline fragrans (L.) Planchon, Fl. Serres 6: 11, 132, 136 (1851)
    Dracaena steudneri Engl. var. *kilimandscharica* Engl. in P.O.A. C: 143 (1895); T.T.C.L.: 21 (1949). Type: Tanzania, Moshi District: Marangu, *Volkens* 1416 (B, holo.; G, K!, iso.)
    D. smithii Hook.f. in Bot. Mag.: t. 6169 (1875); Baker in F.T.A. 7: 440 (1898); F.W.T.A. ed. 2, 3 (1): 156 (1968). Type: specimen cultivated in Kew Gardens, Jan. 1874 (K!, lecto., chosen by Bos)
    D. ugandensis Baker in F.T.A. 7: 445 (1898). Type: Uganda, "common in hedges", *Scott Elliot* 7264 (K!, holo.)
    D. deremensis Engl. in E.J. 32: 95 (1902); T.T.C.L.: 20 (1949); K.T.S.L.: 639 (1994). Type: Tanzania, Lushoto District, Usambara, Handei, Nguelo, *Scheffler* 67 [protologue says 66?] (B, holo., K!, photo)

NOTE. *D. deremensis* var. *warneckei* of T.T.C.L. : 20 (1949) is a cultivar of *D. fragrans* (L.) Ker-Gawl.; see Bos *et al.* in Edin. J. Bot. 49(3): 311–331 (1992).

UNCERTAIN SPECIES

Dracaena fischeri *Baker* in E.J. 15: 477 (1898) & F.T.A. 7: 441 (1898); T.T.C.L.: 20 (1949). Type: Tanzania, Singida District, Usuri, *Fischer* 589 (B†, holo.)

The type has not been seen – it has probably been destroyed. The description of this species in F.T.A. indicates that the author did not see the leaves. Indeed he describes the inflorescence and flowers, but this description is not adequate to infer this plant's identity.

EXCLUDED SPECIES

Dracaena hanningtoni Baker in F.T.A. 7: 438 (1898); T.T.C.L.: 21 (1949). Type: Tanzania, Shinyanga District: Msalala, *Hannington* s.n. (K!, holo.)

Protologue: "nearly allied to *D. ombet.* Panicle glabrous; pedicels densely clustered, 2–3 mm; perianth cylindrical, 12 mm; tube half as long as the segments; stamens as long as the perianth; style distinctly exserted". The type consists of a (partial?) inflorescence some 22 cm long and unbranched. It looks like it might be a branch of something bigger. There are little clusters of pedicels subtended by triangular bracts 1.2–2 mm long and acute; the pedicels are 1–2 mm long. In the envelope attached to the sheet there is a flower with a tube 5.5 mm long and lobes 8.3 mm long and ± 1.5 mm wide; the style is 17 mm long. There are no other parts present.

TANZANIA. Shinyanga District: Msalala, without date, *Hannington* s.n.
DISTR. **T** 1; only known from the type

The material is not enough to decide the specimen's generic identity. From the degree of exsertion of the stigma, and the appearance of the pedicels, it could be that this material is a partial inflorescence of *Sansevieria ehrenbergii* Schweinf.

Dracaena elliptica Thunb. & Dalm.
Scott Elliot 7988 (K) from Toro, Uganda, cited by Baker in F.T.A. 7: 446 (1898) as *D. elliptica* Thunb. & Dalm. is a specimen of *D. laxissima* Engl.

2. SANSEVIERIA

Thunb., Prodr. Pl. Cap. 5 (29): 65 (1794), **nom. cons.**; N.E.Brown in K.B. 1915: 185–260 (1915); Newton in Eggli, Illustr. Handbook Succulent Pl.: Monocot.: 261–272 (2001)

Sansevieria has been variously spelled as *Sanseviera, Sanseverinia* or *Sanseverina* in early publications, but the name *Sansevieria* was conserved in 1904 by Harms (Marais, 1973).

Acaulescent to shrubby, evergreen, fleshy, very drought-hardy perennials; strongly rhizomatous and forming colonies; glabrous in all parts. Stem absent or present and to 1.5 m high; rhizomes subterranean, on surface or aerial, terete, thick, fibrous, articulated, bearing strongly clasping catapylls. Roots sometimes produced from undersurface of aerial shoots. Leaves sessile, rosulate, distichous or spiral, one to many, flat, crescentic to cylindrical, most with a groove on adaxial side; concolorous or with irregular transverse colour bands; lamina edge fibrous, variously coloured; apex subulate, horny, sharp or with a fragile spine; juvenile leaves flat or crescentic (even for cylindrical-leafed species), thin to thick, succulent. Inflorescence a spike-like raceme, a panicle or corymbose-capitate, usually terminal, but a few lateral. Flowers 1–2 or more per cluster, subsessile, in irregular clusters along the scape, opening haphazardly above and below, opening towards evening and for one night only, scented; with one or two major membraneous or thin scaly bract(s) subtending each cluster, one per flower. Perianth greenish white to pinkish green; tube cylindric, glabrous; segments 6, subequal, spreading, linear; stamens 6, inserted at the base of the perianth-lobes; filaments filiform, usually longer than the lobes; anthers versatile, linear-oblong, dorsifixed, dehiscing longitudinally; ovary superior, free, globose, 3-celled; ovules erect, solitary in each of the three chambers; style long, filiform; stigma capitate. Fruit a berry, membranous, soon bursting; seeds 1–3, globose; testa lax, fleshy; embryo straight, placed near the base of the albumen.

A genus of about 50 species from Africa, Madagascar and Arabia.

4. Plants with a small erect stem at least 2.5 cm long 5
   Plants acaulescent 6
5. Leaves banded, with soft tip, stem with aerial branching . .   1. *S. parva*
   Mature leaves not banded, slightly horny at tip, stem with
      subterranean branching . . . .   2. *S. grandicuspis*
6. Leaf 2.5–7 cm wide, without a petiole; cultivated . . . .   *S. trifasciata*
   Leaf less than 1.5 cm wide, with distinct petiole 7
7. Leaf shape lanceolate-spathulate; petiole 6–15 cm long . .   3. *S. subspicata*
   Leaves sword shaped, petiole 30–60 cm long . . . .   4. *S. nilotica*
8. Inflorescence more than 450 cm long (cultivated) . . . . . .   *S. longiflora*
   Inflorescence less than 400 cm long 9
9. Bracts more than 7 cm long . . . .   5. *S. bracteata*
   Bracts less than 7 cm long 10
10. Inflorescence less than 30 cm long . . . .   6. *S. nitida*
    Inflorescence more than 30 cm long 11
11. Plant with surface branching on shallow soils . . . .   9. *S. conspicua*
    Plant with subterranean branching 12
12. Leaves without banding 13
    Leaves banded at least on the lower surface 15
13. Plants with 1–3 leaves, petiolate; leaves with reddish brown
       margin . . . .   11. *S. dawei*
    Plants with 2–8 leaves, no distinct petiole; leaves with reddish
       brown or greyish white margin 14
14. Rhizome to 50 mm thick; leaves 4–8, margins chestnut brown   8. *S. frequens*
    Rhizome ± 20 mm thick; leaves 2–8, margins greyish white   10. *S. forskaoliana*
15. Leaves 1–2; inflorescence 90–130 cm long . . . .   14. *S. raffillii*
    Leaves 2–8; inflorescence 40–75 cm long 16
16. Rhizome 25–38 mm thick; leaves 8.8–17 cm wide . . . .   12. *S. grandis*
    Rhizome 12–25 mm thick; leaves less than 8 cm wide 17
17. Leaves 0.8–1.2 cm wide . . . .   7. *S. aethiopica*
    Leaves 4–8 cm wide . . . .   13. *S. hyacinthoides*
18. Mature leaf blade midsection circular or thicker than wide
       (at least the top half of the leaf), the channel width
       should be less than 25% of the circumference of the leaf
       cross-section when the leaf was fully turgid 19
    Mature leaf blade semicircular, V- or crescent-shaped, 1.5
       to 5 times as wide as thick 23
19. Plants with subterranean branching . . . .   15. *S. volkensii*
    Plants with aerial branching 20
20. Leaves compact, 10–40, straight, in whorls of 5 . . . .   16. *S. francisii*
    Leaves open, 5–12, recurved, not in compact whorls 21
21. Leaves 6 12 cm long, spine 5–7 cm long . . . .   18. *S. ballyi*
    Leaves more than 14 cm long, spine less than 1 cm long 22
22. Stem 0.6–1 cm diameter; bracts 2–4 cm long . . . .   17. *S. gracilis*
    Stem 2–3.5 cm diameter; bracts 4.4–6.9 cm long . . . .   19. *S. suffruticosa*
23. Stem with aerial branching . . . .   20. *S. bella*
    Stem with subterranean branching 24
24. Pedicels 5–8 cm long; rhizome less than 1.5 cm diameter   21. *S. deserti*
    Pedicels less than 2 cm long; rhizome more than 1.5 cm
       diameter 25
25. Leaves numerous (7–14) with numerous deep grooves, light
       grey green, basal margins brownish . . . .   22. *S. sordida*
    Leaves few (2–5), without obvious vertical grooves, greenish,
       basal margins dark-green . . . .   23. *S. patens*

CULTIVATED TAXA

Sansevieria longiflora Sims

Perennial acaulescent herb with subterranean rhizome. Leaves spreading, (3–)4–6, linear-acuminate to oblanceolate, 20–55 × (4–)7.5–8.5 cm, flat but somewhat concave near base, coriaceous, smooth, irregularly blotched with darker and lighter transverse grey bands which become slightly faded with age, margin entire, reddish or yellowish with whitish grey membrane; petiole base almost 25 × 2–3 cm. Inflorescence an erect dense ovoid raceme, (300–)450–600 cm long; flowering part 15–40 cm, with flowers overlapping, flowers (2–)3–4(–6) per cluster; bracts (15–)22–25 mm long, brownish-green, membranous; pedicels 4–5 mm long, 2–3 mm wide, jointed close at the base. Flower perianth 33–40 mm long; tube greenish-white, 8–10 × 2 mm; lobes white, linear, 25–30 mm long, obtuse, revolute; anthers 4 mm long. Fruit a globose berry.

Originally from southern Africa; cultivated at Amani, and possibly gone wild in forest margins on the Korogwe–Handeni road, 7 Apr. 1954, *Faulkner* 1387!

Sansevieria trifasciata Prain

Acaulescent herb with subterranean rhizome. Leaves very erect, 1–2(–6), linear-lanceolate, 30–120 × 2.5–7 cm, banded on both surfaces from base to apex, very distinct light dull green or clear whitish-green and deep grass-green to almost blackish-green, overspread with slight glaucous bloom, margins narrow, grey-green (var. *trifasciata*) or golden yellow (var. *laurentii*). Inflorescence (30–)40–75 cm high, axes 0.3–0.8 cm thick; flowers (1–)3–4 per cluster; pedicels 0.5–0.8 cm long, pedicels upward-curved. Flowers pale greenish or greenish-white, perianth 1.9–2.3 cm long. Fruit a bright orange globose berry, 0.8 cm diameter, 1–3-seeded.

var. *trifasciata* has been cultivated in Kenya, Nairobi: Eastleigh, St. Theresas Catholic Church, 12 Dec. 1971, *Mwangangi & Kasyoki* 1891a!

var. *laurentii* N.E.Br. has been cultivated in Kenya, Nairobi: Eastleigh near St Theresa Catholic Church., 12 Dec. 1971, *Mwangangi & Kasyoki* 1891b!

NATIVE TAXA

1. **Sansevieria parva** *N.E.Br.* in K.B. 1915: 233, fig. 13c–f (1915); Mbugua in U.K.W.F. ed. 2: 312 (1994); L.E. Newton in Ill. Handbook Succ. Plants 1: 268 (2001). Type: Kenya, Naivasha District: near Gilgil River, 1906, *Powell* 15 (K!, holo.)

Rosulate herb with subsurface to aerial branching; rhizome brownish orange, (1–)1.5–2 cm thick; stem (1–) 2.5–8 cm long, 0.5–0.8 cm in diameter, sometimes concealed completely between the leaves. Leaves ascending or suberect, slightly recurved-spreading at the upper part, linear-lanceolate to lanceolate, (6–)8–14(–18) per plant, 20–47 × 0.8–1.4 cm, shallowly concave and folded longitudinally, rounded or obtusely keeled on the back, smooth on both surfaces, tapering at the apex into a stout subulate tip (2–)3.5–5 cm long with soft green point, base narrowed into a petiole up to 5–7 cm long, flat on the face and very rounded at the back but broadly clasping at the base; young leaves very clearly banded on both surfaces with transverse bands of dark bright green and paler green which fade with age hence deep green, margins green, soft; basal scaly leaves 3–4 per plant, 1.5–5.2 × 2.2 cm, membranous, acute, subulate tip 0.3 cm long, grey-green, margin narrow, with evident vertical veins. Inflorescence a lax raceme (25–)30–50(–60) cm long; axis greenish-purple to pale green, 0.2–0.3 cm wide, smooth, glaucous; flower clusters on upper third of axis, flowers (1–)2 per cluster, always solitary in lower and upper parts of the inflorescence, ascending; lower inflorescence bracts 3–4(–5), 1.5–2.5 cm long, the basal clasping, membranous, acute; bracteoles spreading, narrowly lanceolate, 3–4(–6) mm long, membranous, acute, each subtending a flower; pedicel 2–3 mm long, the base hardly jointed. Flower perianth (18–)21–23 mm long; tube pale pinkish-white, 9–12 mm long, swollen at base, lobes greenish grey at base, grey-white in middle, pinkish-brown at apex, linear, 9.5–11 mm long, obtuse; filament 15–17 mm long, anthers 1.5 mm long; ovary greenish grey, elongate, (1.5–)2.4–2.7 mm wide. Fruit a yellow berry, 5–8 mm in diameter.

UGANDA. Kigezi District: Obugwegwe (Lukiga) Kamwezi, Aug. 1949, *Purseglove* 3095!; Masaka District: Masaka–Mbarara road, ± 3 km towards Mbirizi, 30 Aug. 1977, *Pfennig* 1244!; Mengo District: Kipayo, 20 Nov. 1915, *Dummer* 2639!
KENYA. Kiambu District: East Kedong Escarpment, 25 Jan. 1974, *Brandham* & *Cutler* 4/30!; Masai District: Narok, 21 Sep. 1992, *Mbugua* 358!; Teita District: Tsavo National Park, Mazinga Hill near Voi, 26 Mar. 1975, *Bally* 13416!
DISTR. U 2–4; K 3, 4, 6, 7; not known elsewhere
HAB. Dry forest, near bases of termite hills, rocky sites in bushland, also in woodland; 600–2400 m
USES. Used as a fodder by farmers during droughts around Kijabe area – Kenya.
CONSERVATION NOTES. Least concern

SYN. *S. dooneri* N.E.Br. in K.B. 1915: 231, fig. 13 (1915); L.E. Newton in Ill. Handbook Succ. Plants 1: 264 (2001). Type: Kenya, Naivasha/Masai District: Rift Valley, near the Kedong River, *Dooner* s.n. (K!, holo.), **syn. nov.**

NOTE. During fieldwork it was found that leaf banding varies from very bright and evident markings to faint or no bands on the plants growing in the shade. Leaf size also varies much, depending on a number of factors, such as water availability, soil type, etc.

In his revision of this genus Brown (1915) stated that *S. dooneri* was closely related to *S. parva*, though distinguishing it by "its less evident stem and less erect habit, the leaves being much more recurved; they do not have a very distinct petiole, and their subulate points are usually much shorter; also the colour is of a much darker and dull green, with very inconspicuous pale markings". The two taxa, however, grow together in their natural habitats. The type localities are not very far apart i.e. near Gilgil River and in the Kedong Valley – within 30–40 km of each other and both in the Rift Valley. After careful study and comparison, it has been decided that the two are actually one species exhibiting variation due to microhabitat factors.

2. **Sansevieria grandicuspis** *Haw.*, Synop. Pl. Succ.: 67 (1812), as *S. grandiscuspis*; Baker in F.T.A. 7: 336 (1898); Gérôme & Labroy in Bull. Mus. Hist. Nat. Paris 1903: 172, fig. 13 (1903); N.E. Brown in K.B. 1915: 231 (1915); L.E. Newton in Ill. Handbook Succ. Plants 1: 266 (2001). Type: none indicated

Rosulate herb; rhizome subterranean or on surface, greyish brown, 0.5–1 cm thick; stem 3.1 cm long, 0.8 cm in diameter, probably not always evident above the ground. Leaves erect to suberect, ascending-spreading, 5–9(–15), linear-lanceolate, 17.5–50(–55) × 1.3–2.5(–3.8) cm, 0.3–0.4 cm thick, smooth, channel on inner surface deepest at base to flat at apex, subulate tip, 1.7–5(–6) cm long, green, horny, faintly banded on both surfaces but obliterated on maturity, with 5–7 longitudinal shallow furrows on the back, margins green but whitish with age or injury; petiole 5–15 cm long, but no distinct point of petiole emerging. Inflorescence a spike-like raceme ± 38 cm long, axes greenish or light yellow, 0.2–0.3 cm across, flowers scattered along the apical 10–12 cm only, 1–2 per cluster, erect to suberect; lower inflorescence bracts 3, greenish-light brown, lanceolate to filiform, 1–13.5 cm long, 5–10 cm apart; bracteoles 0.3–0.5 cm long, membranous, acute, subtending each flower; pedicels 0.2–0.4 cm long, very thin. Flowers greyish green, 1.3–1.9 cm long; tube 0.5–0.8 cm long, lobes 0.8–1.1 cm long, fully backward rolled when fully open; filaments 1.1–1.3 cm long, anthers 0.2 cm long; style 2–2.5 cm long. Fruit a globose berry.

TANZANIA. Arusha District: 14 km E of Arusha, Lake Duluti, Aug. 1965, *Beesley* 165!
DISTR. **T** 2; not known elsewhere
HAB. Lake banks of volcanic tuff; ± 1420 m
USES. No data
CONSERVATION NOTES. Data deficient (DD)

SYN. *S. ensifolia*, Haw. Synop. Pl. Succ.: 66 (1812). Type: none indicated
    S. pumila, Haw. Synop. Pl Succ.: 67 (1812). Type: none indicated

NOTE. The inflorescence is described here for the first time.
    A species that resembles *S. parva* but differs in having no bands at all, a very long petiole, an acute extended subulate leaf-tip with a hardened apex, larger leaf size and very long narrow lower inflorescence bracts on the inflorescence. This species differs from *S. zeylanica* in having a long leaf-tip, no leaf bands and long lanceolate to filiform lower inflorescence bracts.

3. **Sansevieria subspicata** *Baker* in Gard. Chron. 1889: 436 (1889); N.E. Brown in K.B. 1915: 234 (1915); L.E. Newton in Ill. Handb. Succ. Pl. 1: 271 (2001). Type: Plant cultivated and flowered in 1889 at Kew from material sent in 1866 from Mozambique, Delagoa Bay, by *Monteiro* s.n. (K!, holo.)

Acaulescent herb; rhizome 0.5–1 cm thick, with numerous subsurface roots and scaly membranous leaves at leaf bases. Leaves ascending-spreading and slightly recurving, (3–)4–7(–9) per plant, narrowly lanceolate, (15–)23–27(–32) × (2.5–)3–5(–6) cm, 0.2 cm thick, coriaceous, smooth, acute, subulate tip green, 0.7–1.9 cm long, leaf blade somewhat folded longitudinally, the margins green, both surfaces distinctly banded but clearer on under surface; petiole (6–)10–12(–15) cm long, 0.3–0.4(–0.6) cm wide, bases appear greyish. Inflorescence racemose, 15–30(–42) cm long, axes 0.3 cm wide, flowers on upper half of axis, 1–2 per cluster; lower inflorescence bracts 4(–6), lanceolate, 1.2–2.6 cm long, acuminate, dotted with purple; bracteoles spreading, lanceolate, 0.1–0.2 cm long, acuminate, membranous; persistent portion of pedicels 0.3–0.4 cm long, jointed close under the flower. Flower tube cylindric, 4.4 cm long, swollen at base, lobes whitish-green, 2 cm long, 0.2 cm wide, wider near the apex, slightly tapering to the base. Fruit a globose berry.

var. **concinna** (*N.E.Br.*) *Mbugua* stat. & comb. nov. Type: Mozambique, Beira, *Dawe* 1 (K!, holo.)

Rhizome 0.5 cm thick; scaly membranous leaves at leaf bases, much smaller in size than in var. *subspicata*. Leaves ascending-spreading and slightly recurving, narrowly lanceolate, (15–)23–27(–32) × 2–3(–5) cm, acute.

TANZANIA. Rufiji District: Misikitini, 12 Aug. 1937, *Greenway* 5060!
DISTR. **T** 6; Mozambique, South Africa
HAB. Very locally dominant with two other species in dense shade of *Xylocarpus, Tamarindus, Maulikara, Sideroxylon* fringing bush on a cliff top; near sea level
USES. No data
CONSERVATION NOTES. Least concern

SYN. *S. concinna* N.E.Br. in K.B. 1915: 233, fig. 14 (1915); L.E. Newton in Ill. Handbook Succ. Plants 1: 263 (2001)

NOTE. *Sansevieria subspicata* var. *subspicata* has been found only in Mozambique.
This species has much longer flowers than *S. parva*. The leaves are coriaceous, shiny, erect and longer, with a deeper petiole channel and no surface rhizomes.

4. **Sansevieria nilotica** Baker in J.L.S. 14: 548 (1875); N.E. Brown in K.B. 1915: 238 (1915); Gurke in P.O.A. B.: 367, t. 5, fig. J (1895); Baker in F.T.A. 7: 332 (1898); Demel in Fl. Eth. & Eritr. 6: 82 (1997); L.E. Newton in Ill. Handbook Succ. Plants 1: 268 (2001). Type: Sudan, White Nile banks 'in ditione Muro', 29 July 1863, *Petherick* s.n. (K, holo.)

Acaulescent, rosulate herb; subterranean rhizomes greyish brown, 1.2 cm thick. Leaves 2–3, sword-shaped, (50–)100–120(–130) × 2.5–3.2(–5.5) cm, 0.1 cm thick, above the base 0.4 cm thick, with a long channel, flexible but tough, fairly flat, smooth, the sides of the central part quite parallel for 30 cm or more, narrowing at the apical part into a soft green subulate tip 0.3–0.5 cm long, margin greyish green; whole leaf conspicuously marked with very many closely placed irregular zigzag transverse narrow bars of dark green and pale green, but most of the markings disappear with age and on herbarium material; petiole 30–40(–60) cm long; scaly basal leaves 4–5, 2.5–18 cm long, acute, subulate tip 0.6–3.5 cm long, membranous, greyish white, parallel veined visible on both surfaces, ensheathing the peduncle. Inflorescence a lax raceme with irregularly scattered flower-clusters, 37–45 cm long, 0.4–0.5 cm across; flowers 2–9 per cluster, 6–9 at lower nodes, 2–3 at upper nodes; lower inflorescence bracts ± 5, ovate-lanceolate, 0.1–7.5 cm long, clasping the lower portion, the apex tapering into a long subulate point 0.4–1 cm long or acute; bracteoles lanceolate, scarious, the lower longer than pedicels, the upper shorter; pedicels (0.4–)0.5–0.6(–1) cm long, articulated above the middle. Perianth whitish green, 2.1–2.5 cm long, tube 0.9–1 cm long, lobes 1.2–1.5 cm long; style whitish grey, 2.4–2.9 cm long, exserted. Fruit a yellowish berry, 0.7–0.9 cm wide, 1–2-seeded.

UGANDA. Ankole District: Ruizi River, 6 Oct. 1950, *Jarrett* 197!; Mengo District: near Jumba [Fumbua], Jan. 1917, *Dummer* 3154! & Kampala–Bale road, Bukoloto, 21 Sep. 1973, *Pfennig* 1221!
DISTR. **U** 2, 4; Central African Republic, Sudan, Ethiopia
HAB. Open woodland; 1200–1400 m
USES. No data
CONSERVATION NOTES. Least concern

SYN. *S. nilotica* Baker var. *obscura* N.E.Br. in K.B. 1915: 238 (1915); L.E. Newton in Ill. Handbook Succ. Plants 1: 268 (2001). Type: flowered at Kew in June 1913 from a plant sent from Uganda, without precise locality, by *Dawe* s.n. (K!, holo.), **syn. nov.**
Acyntha massae Chiov. in Atti R. Acad. Ital. Mem. 11: 58 (1940). Type: Ethiopia, Jimma, *Massa* 209 (FT, holo.)
Sansevieria massae (Chiov.) Cufod. in E.P.A.: 1570 (1971)

USES. This is a species easily confused with *S. trifasciata*, since it also has a green margin and the leaf sizes are similar. Distinguishing characters, however, include the leaves being generally narrower, light brownish green, almost glossy on live material with parallel veins and growth marks visible, and very faded or no bands; the inflorescence and the pedicels are shorter and not recurved.

The variety *S. nilotica* var. *obscura* N.E.Br. can not be recognised as a separate taxon using the characters stated in the protologue and observed on the type specimen. It is probable that the more glossy appearance and larger size of live material could be attributed to a great extent to environmental factors.

5. **Sansevieria bracteata** *Baker* in Trans. Linn. Soc. ser. 2, 1: 253 (1880); N.E. Brown in K.B. 1915: 257 (1915); Baker in F.T.A. 7: 333 (1898); Hiern, Cat. Afr. Pl. Welw. 2: 25 (1899); T.T.C.L.: 22 (1949); L.E. Newton in Ill. Handbook Succ. Plants 1: 262 (2001). Type: Angola, Pungo Andongo, *Welwitsch* 3751 (BM!, holo.; K!, P!, iso.)

Acaulescent herb; rhizome orange-vermilion outside, white inside, 1.7–2.3 cm thick. Leaves erect, rigid, 4–6 per plant, lanceolate, 37–67 × (4–)5–7.5 cm, medium to thick, tapering into subulate tip 0.5 cm long, tapering from below the middle into a stout concave-channelled petiole, keeled below, margins brownish red, 1 mm wide, hardened, greenish grey membrane, sometimes with whitish edge where damaged; variegated on both surfaces with irregular zigzag pale green or whitish green bands which are usually much broader than the dark-green ones alternating with them, very faint on old specimens, basal bands 2.5–3.3 cm wide, the rest ± half as broad, closely placed and more or less spotted, slightly rough both surfaces; veins faintly visible on both surfaces; 3–4 basal scaly leaves, ovate to oblanceolate, 10–14 × 3–4 cm, longitudinal veins evident, margins grey-white, membranous. Inflorescence a spike-like raceme, erect to semi-erect, dense, 45–60 cm long, axes 0.6–1 cm thick; flowers 3–5 per cluster; lower inflorescence bracts 5–7, 4.5–6 cm apart, ovate, 7–10 cm long, acute, membranous; bracteoles greenish white, lanceolate, 3.3–3.7 × 0.5–0.7 cm, membranous, subtending each cluster; pedicels 0.4–0.6 cm long, persistent part 0.2–0.3 cm long, jointed above the middle. Flowers white-green, erect; perianth tube 8–10 cm long, lobes linear-spathulate, 2.5–3.3 cm long, 0.3–0.4 cm wide; filaments 9 cm long, anthers 0.3–0.4(–0.5) cm long; style 9–10.5 cm long. Fruit a globose berry.

TANZANIA. Mwanza District, 23 Feb. 1933, *Wallace* 614!; Morogoro District: Uluguru Mts, 11 Feb. 1933, *Schlieben* 3632!; Ulanga District: Mbangala station, 20 Feb. 1932, *Schlieben* 1801!
DISTR. **T** 1, 6; Zambia; Angola
HAB. Few data – riverine forest?; 600–1350 m
USES. None recorded for our area
CONSERVATION NOTES. Possibly Least Concern

SYN. *S. aubrytiana* Carriere in Rev. Hort. 1861: 448 (1861), *non* Gérôme & Labroy (1903)

NOTE. This species has the longest bracts of the whole genus, and can be only be confused with *S. longiflora* N.E.Br., especially because the leaves are almost the same size; it differs in the bracts, these are 3.3–3.7 cm long as opposed to 2.2–2.5 cm in *S. bracteata*. The inflorescence is also more crowded and longer.

6. **Sansevieria nitida** *B.J.Chahinian* in Cact. Succ. J. U.S.A. 73(3): 120, figs. 1–3 (2001). Type: Kenya, Kitui/Kilifi District: Galana Ranch, *Chahinian* 301 (MO, holo.; NY, iso.)

Acaulescent herb; rhizome creeping, orange, sturdy, to 2 cm in diameter. Leaves 1–5, usually 2, erect-spreading and sometimes twisting, lanceolate, to 40 × 6 cm, 2 mm thick, apex acute, base narrowed into a pseudopetiole with keel 4 mm thick, upper surface smooth and shiny, lower surface rough and dull, dark green with dense lighter grey-green blotches cross-banding, margins fibrous and chestnut brown, the edges withered. Inflorescence simple, spike-like, 15–24 cm long, with dense flowers in upper $^3/_4$, flowers 1(–2) per cluster, axis 5 mm in diameter, light green with faint short longitudinal lines; basal bracts 4, triangular, 11–20 × 4–11 mm; lower inflorescence bracts lanceolate, lower 10 × 5 mm, upper ones 3 mm long; pedicel 2 mm long. Flowers greenish white with purple lines; perianth tube 18–24 mm long, inflated at base, lobes convoluted, 20–26 mm long; filaments 29 mm long, anthers 3 mm long; style 50–54 mm long.

KENYA. Kitui/Kilifi District: Galana Ranch, *Chahinian* 301
DISTR. **K** 4/7; not known elsewhere
HAB. No data
USES. No data
CONSERVATION NOTES. Data deficient

NOTE. Closest to *S. conspicua* but differs in fewer and shorter leaves, a shorter scape, shorter tube and lobes, and 1(–2) flower(s) per cluster.

7. **Sansevieria aethiopica** *Thunb.*, Prodr. Pl. Cap.: 65 (1794); N.E. Brown in Bot. Mag., 139: t. 8487 (1913) & in K.B. 1915: 230 (1915); F.S.A.: 5 (3): 7 (1992); L.E. Newton in Ill. Handbook Succ. Plants 1: 261 (2001). Type: South Africa, *Thunberg* s.n. (UPS)

Acaulescent rosulate herb; rhizome subterranean, 1.2 cm thick, with orange-grey roots. Leaves erect to slightly recurved, half folded, 2 per plant, lanceolate, 25–32 × 0.8–1.2 cm, acute, subulate tip 0.3–0.5 cm long and slightly horny, margin bright brownish red, less than 1 mm wide, grey-green bands against dark green background on both surfaces, bright brownish red at base becoming grey-brown towards apex; sheathing base turns greyish green, smooth on both surfaces; basal scaly leaves 2, 1.9–2.6 × 2 cm, membranous, obtuse to acute, evident veins converging at apex. Inflorescence a spike-like raceme 54 cm long, axis 0.3–0.5 cm across; flowers 2–3 per cluster; lower inflorescence bracts 5, 5.5–6.5 cm apart, erect, greenish grey, 2–4.5 × 0.6 cm, acute, membranous; bracteoles greenish grey, 0.5–0.8 cm long, acute, membranous; pedicels 0.2–0.3 cm long, not articulated. Flower perianth 3.9–4.1 cm long, 0.2 cm wide; tube greyish purple to green, 1.8–2.1 cm long, narrowed above the ovary, lobes purplish green, 1.5–2 cm long; stamens 2–2.2 cm long, exserted beyond lobes, anthers 3.5 mm long; style 4–4.3 cm long, exserted beyond the stamens. Fruit a yellowish green berry, 0.5–1.2 cm in diameter.

subsp. **itumea** *Mbugua* **subsp. nov.** a subsp. *aethiopica* differt floribus minus numerosis (tantum 2–3 per fasciculum) et brevioribus, perianthio 3.9–4.1 x 0.2 cm lobis 1.5–2 cm longis tubo 1.8–2.1 cm longo, foliis minus numerosis (2) et minoribus 25–31.5 x 0.8–1.2 cm ubique laevibus differt. Type: Kenya, Tana River District, 30 km south of Galole, *Bally & Smith* B14404A cult. in Germany by *Pfennig* 312 (HBG!, holo.; B, iso.)

Differs from the type by having shorter and fewer flowers per cluster, 2–3 per cluster, perianth 3.9–4.1 × 0.2 cm; lobes 1.5–2 cm long; tube 1.8–2.1 cm long; fewer, leaves 2 per plant, smaller, 25–31.5 × 0.8–1.2 cm, smooth on both surfaces.

KENYA. Tana River District: 30 km south of Galole, 26 Feb. 1971, *Bally & Smith* B14404A, cultivated in Germany by *Pfennig* 312!
DISTR. **K** 7; only known from the type
HAB. Grows within thickets in dry arid lowlands of the coastal Kenyan region
USES. Berries eaten by birds
CONSERVATION NOTES. Known in only one locality and probably Vulnerable (VU-D2).

NOTE. *Sansevieria aethiopica* Thunb. var. *aethiopica* occurs in Zambia, Zimbabwe, Botswana, Namibia and South Africa.

8. **Sansevieria frequens** *B.J.Chahinian* in Cact. Succ. J. U.S.A. 72(3): 130 (2000); L.E. Newton in Ill. Handbook Succ. Plants 1: 266 (2001). Type: Kenya, North Nyeri District: Ngare Ndare farm, *Chahinian* 785 (MO, holo.; NY, iso.)

Acaulescent herb; rhizome brown, sturdy, creeping, to 5 cm in diameter. Leaves 4–8, erect, oblanceolate, to 90 × 15 cm, stiff, smooth, apex obtuse, formed by withered edge, dull green without cross-banding (except in juvenile stage), margins fibrous and chestnut-brown, the edges detaching into fibres. Inflorescence crowded in upper part,

60–90 cm long; peduncle green, 25 mm in diameter at base; bracteoles 3–4, triangular, withered, to 5 × 4 cm; lower inflorescence bracts 8–35 mm long, 3–25 mm wide; flowers in clusters of 4–6; pedicel 5 mm long. Flowers greenish white; perianth tube 18–20 mm long, inflated at base, lobes convolute, 26–28 mm long; filaments 25 mm long, anthers 3 mm long; style 56–58 mm long. Fruit orange, globose, 4–6 mm; seed bony, 3–5 mm.

Kenya. Trans Nzoia District: NE Elgon, May 1957, *Tweedie* 1441!; Baringo District: Kabarnet and Tambach, 11 June 1980, *Pfennig* 82!; Embu District: Embu–Meru road, ± 2 km N of Ishiara, 18 Mar. 1972, *Pfennig* 1056!
Tanzania. Tanga District: Ukereni, 29 June 1918, *Peter* 24156!; Ufipa District: Lake Tanganyika, Finger Point, 20 Oct. 1964, *Richards* 19204!; Uzaramo District: Observation Hill road, 10 June 1938, *Vaughan* 2817!
Distr. **U** (see note); **K** 3, 4; **T** 3, 4, 6; Ethiopia
Hab. Dry bushland, rocky sites, secondary grassland; 50–2600 m
Uses. Leaf fibre used by spectacled weaver bird for nest-building
Conservation notes. Least concern

Note. Chahinian gives sight records for **U** 2 and **K** 5, 6, and states the species is common in the Eastern Suguta valley, at Rumuruti and Ngare Ndare, in Meru National Park and along the Nairobi–Mombasa and Nairobi–Magadi roads.
   Easily confused with *S. raffillii*, but has wider and shorter as well as a smooth leaf surface. The inflorescence is crowded and brown in colour, especially the peduncle.

   9. **Sansevieria conspicua** *N.E.Br.* in K.B. 1913: 306 (1913) & in K.B. 1915: 243, fig. 19 (1915); Mbugua in U.K.W.F. ed. 2: 312, t. 140 (1994); L.E. Newton in Ill. Handbook Succ. Plants 1: 263 (2001). Type: plant flowered at Kew in Sep. 1909, sent in 1906 from Kenya, Kilifi District: near Mazeras, *Powell* 12 (K!, holo.)

Acaulescent herb, sub-erect to procumbent; rhizome subterranean, surface branching on shallow soils, reddish or purplish, 1.7–3.7 cm thick. Leaves ascending-spreading, 3–5(–6), lanceolate or elongated lanceolate, 22–75 × 5–7.5(–8.1) cm, 0.2 cm thick at middle, acute, very smooth, apex 0.2–0.3 cm long, narrowed from below the middle to the base, margin brownish red, 1 mm broad, slightly twisted and wavy, hardened, narrowly edged with white; both surfaces dull green, slightly darker above, numerous longitudinal lines of darker green from base to apex, very faint bands on under surface that disappear with age, apex almost obtuse to retuse when subulate tip is broken off; petiole indistinct; basal scaly leaves 3.5–8 × 3.5 cm, some with yellowish brown margin, others grey-white membranous, vertical veins evident. Inflorescence a spike-like raceme, 30–60(–90) cm long, axis greyish green, suffused with dull purple marked with minute paler specks, 0.6–0.8 cm across; flowers 1–2(–3) per cluster; lower inflorescence bracts 4–5, pale brownish, 3.1–8.8 cm long, submembranous, tapering to a very acute point; bracteoles reflexed linear-lanceolate, 0.5–1.3 cm long, acute, membranous, remain attached on inflorescence even after flowers fall off; pedicels 0.4–0.6(–0.8) cm (0.2–0.5 cm long when dry), many pedicels thick, 1.5–2 cm, articulated, no deciduous part. Flowers 3.5–4 cm long; tube greenish white, 0.3–0.6 cm long, rather slender, swollen at base; lobes white, linear, 3.2–3.4 cm long, revolute, obtuse. Fruit a globose berry 0.5–0.7 cm in diameter.

Kenya. Machakos District: Kibwezi, 9 Mar. 1982, *Pfennig* 88!; Kilifi District: Sabaki R. 8 km N of Malindi, 31 Oct. 1961, *Polhill & Paulo* 682!; Kwale District: Buda Mafisini Forest, 17 Aug. 1953, *Drummond & Hemsley* 3845!
Tanzania. Shinyanga District: Shinyanga, Apr. 1935, *Burtt* 5213!; Lushoto District: Mashewa–Magoma road, 8 km SW of Mashewa, 8 July 1953, *Drummond & Hemsley* 3228!; Uzaramo District: Kunduchi, 16 km NNW of Dar es Salaam, 20 Feb. 1971, *Harris & Tadros* 5678!; Zanzibar, Massazini / Chukwani / Chwaka, 31 Aug. 1959, *Faulkner* 2341!
Distr. **K** 4, 7; **T** 1, 3, 6, 7; **Z**, **P**; Zambia, Malawi and Mozambique
Hab. Littoral forest/evergreen bushland, also in crevices and earth pockets in coral, often locally common; inland more rare, on rocky outcrops and in riverine thicket; 0–1200 m
Uses. No data

Conservation notes. Least Concern

Note. The specimens along the east coast of Africa have suberect leaves, smooth, no bands and are generally much shorter than the inland specimens. The specimens from Malawi have longer and more stiff leaves. The suberect to procumbent growth form is one of the major diagnostic characters.

10. **Sansevieria forskaoliana** (*Schult.f.*) *Hepper & Wood* in K.B. 38(1): 83 (1983); Thulin, Flora Som. 4: 30 (1995); Demel in Fl. Eth. & Eritr. 6: 82 (1997); L.E. Newton in Ill. Handbook Succ. Plants 1: 265 (2001). Type: Yemen, Al Hadiyah [Hadie], *Forsskål* 9 (C, holo.)

Acaulescent herb; rhizomes ± 20 mm thick. Leaves rosulate, erect to subrigid, (2–)3–6(–8) per plant, lanceolate, 65 × (5–)6.5–7.5 cm, acute with a hardened apical tip 0.2 cm long (dried up in some specimens), margins 1–2 mm wide, sinuate, with 2 orange to brown stripes alongside very narrow grey-white membrane along the margin, rough on adaxial surface but smooth on abaxial surface, flat in the middle to U-shaped in transverse section at base, firm or sub-rigid; obscure or no bands on either surfaces; petiole rudimentary. Inflorescence a terminal or subterminal compact spike-like raceme, (65–)71–76 cm long with 6–7 clusters 1.6–2 cm apart, axis 3–5 mm wide; flowers (4–)5–6(–7) per cluster or fewer towards the apex; lower inflorescence bracts 3–4(–5), 3.5–7 cm long, membranous, acute, with distinct veins; bracteoles greyish brown-white, 1–1.3 × 0.3–0.4 cm, oblanceolate, membranous, acute; pedicels in mature flowers creamy-white, 6.4–9.2 mm long, persistent part 2–4 mm long, jointed at the middle. Flower perianth dark grey-white, 2.6–3.2(–4) cm long, 0.2–0.4 cm wide; tube 2–2.2 cm long, lobes 1.6–1.9 cm long; filaments as long as the lobes, anthers 2.5 × 0.5 mm; style longer than stamens. Fruit a berry, 7–10 × 6–12 mm, 1(–3)-seeded.

Kenya. Northern Frontier District: 10 km from Moyale on Marsabit road, 12 Aug. 1952, *Gillett* 13718!; Machakos District: Nairobi–Voi road near Kibwezi, around Mbuinzau, 19 Feb. 1982, *Pfennig* 1008!; Tana River District: 85 km N of Malindi on road to Garsen, no date, *Robertson* 1778!
Tanzania. Maswa/Masai District: Serengeti National Park, Makoma NW Kopje, 28 June 1974, *Kreulen* 394!; Morogoro District: Nguru Hills, Sep. 1924, *Simmance* 170!; Rungwe District: Kyimbila, 1915, *Stolz* 1597!
Distr. **K** 1, 2, 4, 6, 7; **T** 1, 6, 7; Congo-Kinshasa, Sudan, Ethiopia, Djibouti, Somalia; Yemen
Hab. Dry or evergreen bushland, grassland; 100–1700 m
Uses. Leaf fibre used for string
Conservation notes. Least concern

Syn. *Convallaria racemosa* Forssk., Fl. Aegypt.-Arab.: 73 (1775), *non* L. (1753). Type: Yemen, Hadie, 1763, *Forsskål* 9 (C, holo., microfiche!)
   Smilacina forskaoliana Schult.f., Syst. Veg. 7: 304 (1829), as *forskaliana*
   Sansevieria guineensis (L.) Willd. var. *angustior* Engl. in E.J. 32 : 97 (1902). Type : Ethiopia, Sheik-Hussein, *Ellenbeck* 1242 (B†, holo)
   S. abyssinica N.E.Br. in K.B. 1913: 306 (1913) & in K.B. 1915 : 241 (1915) Type: Ethiopia, near Jana, *Schimper* 1468 (P, holo., K! drawing)
   Acyntha elliptica Chiov., Fl. Somala 2: 421 (1932). Type: plant cultivated in Modena from material sent from Somalia by *Guidotti* (not found)
   Sansevieria abyssinica N.E.Br. var. *angustior* (Engl.) Cufod., E.P.A.: 1569 (1971)
   S. elliptica (Chiov.) Cufod., E.P.A.: 1569 (1971)

Note. A very widespread species, which produces more fruits per inflorescence than any other. Easily confused with *S. raffillii*, but distinguished by having more leaves per plant with a rough adaxial surface, more flowers per cluster, and longer bracts.

11. **Sansevieria dawei** *Stapf* in J.L.S. 37: 529 (1906); N.E. Brown in K.B. 1915: 247 (1915); Mbugua in U.K.W.F. ed. 2: 312 (1994); L.E. Newton in Ill. Handbook Succ. Plants 1: 264 (2001). Type: Uganda, Mengo District: Busiro, *Dawe* 109 (K!, lecto., chosen here)

Acaulescent herb, with a creeping rhizome up to 0.3–0.4 cm thick. Leaves (1–)2–3 per growth, ascending or suberect, elongate-lanceolate, 60–150 × 5.5–11.5(–13) cm, smooth abaxially but slightly rough adaxially, apex 0.4–0.5 cm long, acute, margins reddish brown; dull green and slightly glaucous on both surfaces especially during fast growth, faintly banded, pale green when young; petiole 15–20(–30) cm long, concave-channelled; basal scaly leaves 4–5, 5–16 cm long, longest ones greenish, others purplish blue to green, membranous, grey-green margin. Inflorescence a spike-like raceme with distinct clusters above, 60–90(–120) cm long; flowers 3–4 per cluster; bracteoles greenish or tinged with dull purple, ovate to ovate-oblong, 0.8–1.6 cm long, acute or subobtuse, membranous; pedicels 0.4–0.5 cm long, jointed close under the flower. Flower perianth greyish white, 2.7–3.2(–3.5) cm long; tube 1–2 cm long, lobes 1.6–2 cm long; stamens 2.5 cm long, anthers 0.3 cm long; style 3.1–4.7 cm long, exserted beyond the lobes. Fruit an orange berry, 3–seeded.

UGANDA. Karamoja District: Dodoth County, Kaabong, 17 Sep. 1950, *Dawkins* 642!; Bunyoro District: Lake Albert Flats, Nov. 1939, *Eggeling* 3839!; Mengo District: Kampala, Oct. 1930, *Snowden* 1823!
KENYA. Baringo District: Kabarnet and Tambach, 17 Aug. 1978, *Rauh* 29.113!; Nairobi–Voi road near Kibwezi Forest. 15 May 1971, *Pfennig* 1009!; Teita District: between Voi and the Taita Hills, 31 Mar. 1906, *Grenfell* s.n.!
DISTR. **U** 1, 2, 4; **K** 3, 4, 7; Rwanda, Burundi
HAB. Dry bushland, especially on rocky outcrops, also dry forest; 600–1550 m
USES. Cordage is made from the leaves
CONSERVATION NOTES. Least concern

NOTE. This may be taken for *S. raffillii*, but there are no bands at all on either surface, the plant is larger in size, with a thick peduncle, dark green-bluish leaves (glaucous when young), larger bracts and fewer flowers per cluster.

12. **Sansevieria grandis** *Hook. f.* in Bot. Mag. 129: t. 7877 (1903); N.E. Brown in K.B. 1915: 252 (1915); De Wild., Not. Pl. Utiles Congo: 627, 633 (1903); Holland in K.B. 1907: 369 (1907). Type: Iconotype: t. 7877, Curtis's Bot. Mag. (1903), drawn from specimen sent to Kew from plants cultivated in Cuba

Acaulescent herb, 60 cm long; rhizome subterranean, whitish to green, 2.5–3.8 cm thick. Leaves ascending or ascending-spreading, (3–)4–5, elliptic, stiffly coriaceous, oblong or broadly lanceolate, 30–60(–90) × 8.8–15 (–17) cm, 2–3 mm thick at the middle, flat, narrowed and convolute at base, slightly rough, acute, subulate tip 0.4–0.6 cm long, margins 0.1 cm wide, slightly wavy, hardened, narrow membranous edges when young soon disappearing, dull glaucous-green to somewhat bluish green, transverse bars irregular lighter green on both surfaces but more conspicuous on lower surface; slightly petiolate. Inflorescence 41–60 cm high, axis 0.8–1 cm thick near base, upper two-thirds a compact spike-like raceme; lower inflorescence bracts 4–5, ovate, 1.3–1.9 × 1.3–1.9 cm, obtuse, distant, membranous; bracteoles membranous, ovate or ovate-lanceolate, 0.3–0.6 cm long, acute; pedicels 0.2–0.4 cm long, jointed at apex, with scarcely any deciduous part. Flowers white to grey-white; tube 1.2–1.6(–1.9) cm long, rounded at base, lobes linear, 1.5–1.9 cm long, obtuse, revolute; filaments 1.6–1.8 cm long, anthers 0.3 cm long; style 3.6–4 cm long. Fruit a globose berry; brownish yellow on ripening.

KENYA. Machakos District: Kibwezi, 9 Mar. 1982, *Bally* & *Pfennig* 88!; District unclear: Kericho–Kisumu road ± 10 km from Kapsoit, 23 Mar. 1972, *Pfennig* 1065!; Kwale District: near Samburu Station, 22 Sep. 1961, *Verdcourt* 3220!
TANZANIA. Iringa District: Livingstone Mts, 10⁰ 04'S 34⁰ 34'E, 1 Feb. 1991, *Gereau & Kayombo* 3825!
DISTR. **K** 4, 5, 7; **T** 7; Zimbabwe, South Africa
HAB. Dry bushland or riverine forest; 50–600 m
USES. No data
CONSERVATION NOTES. Least concern

SYN. *S. nobilis* God.-Leb., Sansev. Gigant. Afr. Orient.: 12, name only (1901)

NOTE. This species differs from *S. hyacinthoides* by its wider leaves, shorter flowers, slightly wavy leaf margins, dull glaucous and dark green-blue colour. It may also be taken for *S. conspicua*, but its crowded inflorescence, slightly rough undersurface and shorter pedicels help to distinguish it. Common name: "pendulous *Sansevieria*", because of its growth form in cultivation.
Our material belongs to var. *grandis*.

13. **Sansevieria hyacinthoides** (*L.*) *Druce* in Rep. Bot. Exch. Cl. Brit. Isles 1913 (3): 423 (1914); Obermayer in F.S.A. 5, 3: 5 (1992); L.E. Newton in Ill. Handb. Succ. Plants 1: 266 (2001). Type: Commelin, Praeludia Bot.: t. 33 (1703), iconotype

Acaulescent perennials forming large colonies; rhizome subterranean, 1.2–2.5 cm thick. Leaves rosulate, 2–8, lanceolate to broadly linear, 60–85 × 4–8 mm, flat to somewhat incurved, smooth to slightly rough, subulate tip 0.5 cm long, acute, hard but not horny, margin bright red, entire, 1 mm wide, with a pale membranous edge easily peeling off, with irregular, paler horizontal areas, conspicuous banding may fade with age; keeled towards the base, petiole 20–25 cm long; basal scaly leaves 2–3, 8–12 × 4–6 cm, membranous, apex slightly obtuse, with distinct veins. Inflorescence an erect raceme, 65–70(–75) cm long, axis 0.4–0.6 cm wide; flowers 4–5(–7) per cluster, 2–2.5 cm apart, subtended by 0.4–0.5 cm long membranous bracteoles; pedicels 0.4–0.7(–0.9) cm long. Flower perianth (26–)30–40(–65) mm long; tube 14–20 mm long, lobes cream to pale mauve, 16–20 mm long; filaments ⊥ 2 cm long, thin, anthers yellow, 0.3–0.4 cm long. Fruit orange or yellow, globose, shortly stipitate; seeds globose, ± 8 mm in diameter, the epidermis thick.

TANZANIA. Tabora District: Msisi–Hongasea, no date, *Fischer* 1! & 279!; Rungwe District: Kyimbila, 1915, *Stolz* 1597!
DISTR. **T** 4, 7; Zambia, Malawi, Mozambique, Zimbabwe, South Africa
HAB. no EA data, elsewhere growing in dry bushland or wooded grassland
USES. No data
CONSERVATION NOTES. Least concern

SYN. *Aloe hyacinthoides* L., Sp. Pl. : 321 (1753)
  Aletris hyacinthoides (L.) L., Sp. Pl. ed. 2: 456 (1762)
  Sanseverinia thyrsiflora Petagna in Instit. Bot. 3: 643 (1787), *nom. illegit.* based on *Aloe hyacinthoides* L.
  Sansevieria thyrsiflora (Petagna) Thunb., Prodr. pl. Cap.: 65 (1794) & Fl. Cap.: 329 (1823); Baker in J.L.S. 14: 547 (1875); Baker, Sansevieria 1896: 5 (1896); N.E. Br. in K.B. 1915: 249 (1915)

14. **Sansevieria raffillii** *N.E.Br.* in K.B. 1915: 252, fig. 22 (1915); Mbugua in U.K.W.F. ed. 2: 312 (1994); L.E. Newton in Ill. Handbook Succ. Plants 1: 269 (2001). Type: plant flowered at Kew in December 1910, sent from Kenya, Tsavo district, *Powell* 7 (K!, holo.)

Acaulescent herb; rhizome subterranean, whitish, 3–4 cm thick. Leaves 1–2 per plant, erect to slightly recurved, lanceolate, 65–110(–150) × (4–)5.5–13(–14) cm, 0.8–1.5(–2) cm thick, coriaceous, rigid, smooth but slightly rough, apex 0.3 cm long, semi-fleshy, sometimes broken off, narrowed below the middle into a stout concave petiole 9–12 cm long, margins reddish brown, 1 mm wide, hardened; variegated in juvenile with large elongated oval closely placed blotches or broad irregular transverse yellowish-green bands on a darker green on both surfaces but markings much faded on abaxial surface, more faded with age, slightly glaucous (especially during fast growth); basal scaly leaves up to 8(–10) cm from the base, purplish blue, elongated lanceolate, broadly strap-shaped, acute. Inflorescence a raceme 90–120(–130) cm long, axis 1.5–2.3 cm thick; clusters not in an apparent nodal arrangement, 4.5–6 cm apart, flowers 2–5 per cluster; lower inflorescence bracts 4–5, 4.5–6 cm apart, lanceolate, 5.5–9(–10.5) × 1.2 cm, acuminate, some with subulate apex 1–1.2 cm long, lower inflorescence bracts coloured like the leaves when young and broader, upper thinner and membraneous; bracteoles ovate-lanceolate, 0.5–0.9(–1.2) × 0.2–0.3(–0.5) cm, acuminate, membranous; pedicels 4–5(–7) mm long, jointed above the middle, persistent part 0.3–0.4 cm long. Flower perianth (3.9–)4.5–5.5 cm long; tube greenish white, (1.9–)2.5–2.7 cm long, slightly glaucous, lobes white to grey green, linear, 2–2.5 cm long, obtuse, revolute; filaments 3.8–4.1 cm long, anthers greenish yellow, 0.3 cm long; style 5–5.5 cm long. Fruit a globose orange berry.

UGANDA. Bunyoro: Lake Albert Flats, Butiaba, June 1935, *Eggeling* 2046!
KENYA. Baringo District: Kabarnet–Tambach area, 8 Mar. 1982, *Pfennig* s.n.!; Machakos District: Kibwezi, 25 Mar. 1960, *Bally* 12205!; Kajiado District: Magadi road S of Olorgesaile, 4 Apr. 1969, *Greenway, Napper & Kanuri* 13600!
TANZANIA. Musoma District: North of Lake Magadi on the Seronera track, 20 Apr. 1961, *Greenway & Turner* 10078!; Ufipa District: above Lake Sundu, 8 Nov. 1958, *Richards* 10252; ?Uzaramo [Usalamo], 3 Sep. 1926, *Peter* 44811!; Zanzibar, Kilimani, 8 Aug. 1889, *Stulhmann* 1126d!
DISTR. **U** 2; **K** 3–4, 6–7; **T** 1, 4, 6; **Z**; Sudan, Zambia
HAB. Dry bushland, woodland or dry forest, usually on rocky slopes or near termite hills; may be locally common; (0–)500–2050 m
USES. Useful fibre
CONSERVATION NOTES. Least Concern

SYN. *S. raffillii* N.E.Br. var. *glauca* N.E.Br. in K.B. 1915: 252 (1915); L.E. Newton in Ill. Handbook Succ. Plants 1: 269 (2001). Type: plant flowered at Kew in August 1911, sent from Kenya, Tsavo district, *Powell* 8 (K!, holo.), **syn. nov.**

NOTE. The main distinguishing characters are: 1–2 flowers per cluster, longer flowers, to 4 cm at least; leaves 1–2 per plant , with banded adaxial surface, coriaceous; scaly leaves purplish blue; very long sheaths, to over 8 cm; rhizomes 3–4 cm thick.

15. **Sansevieria volkensii** *Gurke* in P.O.A. C: 144 (1895); Baker in F.T.A. 7: 334 (1898); N.E. Brown in K.B. 1915: 226 (1915); Thulin, Fl. Som. 4: 29 (1995); L.E. Newton in Ill. Handbook Succ. Plants 1: 272 (2001). Types: Tanzania, Lushoto District: Usambara, Rombo, *Holst* 4080 & Moshi District: Lake Chala, *Volkens* 1779 (both B†, syn.). Neotype: Tanzania, Pare District, 3.5 km W of Same railway station, *Wingfield* 1525 (K!, chosen here)

Acaulescent herb; rhizome branching underground, bright orange-red. Leaves 2–7(–8) per plant, mainly erect but older ones ascending to slightly recurving, the outer crescentic with acute edges to the concave channel, the inner cylindric, 50–120(–150) × 1.3–1.9 cm, channeled all down the face, the central leaf with almost indistinct channel, stiff, slightly rough to very rough with a surface like ground glass, gradually tapering to a sharp whitish-brown spine 0.6 cm long, flattish and hardly piercing, the channel much narrower than, or at the basal part sometimes nearly as broad as, the leaf; margins greenish whitish at the basal part, passing into obtusely rounded green edges at the upper part, sharp, with numerous slight grooves or impressed longitudinal lines on the sides and back, dull deep green to bluish green with age, transverse banding decreasing with maturity. Inflorescence a spike-like raceme with clusters in upper part, light greyish green, 25–50(–75) cm high; flowers 3–6 per cluster, very crowded, ascending-spreading; lower inflorescence bracts 2–3, 5–7.5 cm long; bracteoles 0.2–0.3 cm long, membranous, acute; pedicels 0.2 cm long or less, jointed at the apex. Flower tube pale greenish, 1.7–3.2 cm long, 0.3 cm wide, slightly inflated at the base, lobes white or greenish white and minutely dusted with purplish on the back at the apex, linear, 1.3–2 cm long, obtuse, revolute. Fruit a globose berry, 0.6–1.4 cm in diameter.

KENYA. Kitui District: Mutomo Hill, 4 Mar. 1960, *Bally* 12114!; Masai District: Laitokitok [Loitoktok], 21 Mar. 1986, *Pfennig & Rauh* 29S!; Teita District: Mwatate–Voi road, 8 km from Mwatate, 18 Sep. 1953, *Drummond & Hemsley* 4420!

TANZANIA. Moshi District: near Mwanga, 21 Mar. 1986, *Pfennig* 1135!; Pare District: Vudee, 30 Jan. 1930, *Greenway* 2091! & 3.5 km W of Same railway station, 6 Apr. 1971, *Wingfield* 1525!
DISTR. **K** 4, 6, 7; **T** 2–4, 7; Congo-Kinshasa, Somalia
HAB. Dry bushland (*Acacia-Commiphora, Acacia-Euphorbia*), often in rocky sites and may be very common locally; 50–1750 m
USES. Roots used medicinally; leaves squeezed for water by baboons
CONSERVATION NOTES. Least Concern

SYN. *S. intermedia* N.E.Br. in K.B. 1914: 83 (1914) & in K.B. 1915: 211, fig. 6 (1915); Blundell, Wild Fl. E. Afr.: fig. 248 (1987). Type: plant flowered at Kew in Nov. 1913, sent from Kenya, "Tsavo district", *Powell* 9 (K!, lecto, chosen here), **syn. nov.**
S. quarrei De Wild., Contrib. Fl. Katanga, Suppl. 4: 4 (1932). Type: Congo-Kinshasa, Shaba Province, *Quarré* 1261] (BR!, holo.), **syn. nov.**
Acyntha polyrhitis Chiov., Fl. Somala 2: 422 (1932). Type: Somalia, Gelib, *Tozzi* 410 (FT, holo., K, iso.)
Sansevieria polyrhitis (Chiov.) Cufod. in E.P.A.: 1570 (1971)

NOTE. This species closely resembles *S. pearsonii* (Southern Africa), *S. desertii* and a few others within the East African region. It is identified in having a very rough leaf surface – almost like ground glass, and an acute horny spine.

16. **Sansevieria francisii** *B.J.Chahinian* in Sansevieria J. 4(1): 12, fig. (1995); L.E. Newton in Ill. Handbook Succ. Plants 1: 265 (2001). Type: Kenya, Tana River District: Garsen, *Horwood* 432; from cultivated material originally collected in 1982 (MO, holo.; UPS, iso.)

Herbaceous and stoloniferous; stem 10–30 cm long, covered in closely packed leaves in whorls of 5; stolons from stem base and/or from among the leaves. Leaves porrect, cylindrical, 8–15 cm long, slightly rough, tapering to a sharp spiny apex to 5 mm long, upper surface with a flattened shallow channel 25–75% of its length with green margins and with a white membrane over a brown-red longitudinal line reaching halfway; lamina dark green marked with grey-green cross-banding plus 4–6 dark green longitudinal lines turning into grooves, almost to the apex, one of these on the upper surface and through the channel. Inflorescence erect or slightly curved, 12–25 cm long; peduncle 3–8 mm in diameter; flowers dense in the upper half of the inflorescence, 1–2 per cluster; bracts 4, triangular, to 6 × 3 mm; bracteoles absent; pedicel 2 mm long. Flower greenish white to brownish green, perianth tube 16–19 mm long, lobes reflexed, 8–10 mm long; stamens 10 mm long, anthers 2 mm long; style ± 13 mm long.

KENYA. Tana River District: Garsen, no date given, *Horwood* 432
DISTR. **K** 7; not known elsewhere
HAB. No data; altitude low
USES. No data
CONSERVATION NOTES. Data deficient

NOTE. This species resembles *S. gracilis* but differs by the 5-leaved whorls, the compact habit, the short, straight and slightly rough leaves, and the shorter inflorescence.

17. **Sansevieria gracilis** *N.E.Br.* in K.B. 1911: 96 (1911) & in K.B. 1915: 204, fig. 4 (1915); T.T.C.L.: 22 (1949); L.E. Newton in Ill. Handbook Succ. Plants 1: 266 (2001). Type: plant flowered at Kew in 1910, sent from Kenya: Mazeras, *Powell* 11 (K!, holo.)

Herbaceous, procumbent and slightly ascending to spreading (trailing in cultivation); stem absent or up to 30(–90) cm long, 0.6–1(–1.7) cm in diameter, usually 1(–2) branches per plant, sheathed, procumbent to ascending, aerial branching and rooting, rhizome covered with ovate acute spine-pointed tough scaly leaves, the upper turning into leaves, 1–2.5(–4.5) cm long, when dry grey-brown with white membranous margins. Leaves closely packed, rosulate, ascending to recurved,

FIG. 2. *SANSEVIERIA GRACILIS* — **1**, habit, × ²/₃; **2**, inflorescence detail with bract, × 3; **3**, flower, × 2. All from *Robertson* 5445 and *Robertson & Luke* 5371b. Drawn by Juliet Williamson.

(6–)8–12 per plant, inner or fully developed leaves concave near base, cylindric above, 25–75 × 0.6–1.3 cm thick at the top of the sheath, spiral, straight, curved or slightly sinuous, firmly flexible, not rigid, smooth to rough, the outer of each tuft much shorter than the inner, concave down the face, very rounded on the back; sheathing, and concave-channelled for 5–13 cm at the base, gradually tapering to a very acute spine-like brown or whitish point 0.2–0.6 cm long, at first without grooves or channels, becoming faintly to deeply grooved on the oldest or shrivelled leaves, deep green bands on the young leaves and with some slightly darker continuous or interrupted longitudinal lines; margins of the basal sheaths with membranous white edges. Inflorescence a lax spike-like raceme 15–30(–60) cm long, axis light green, 2–4 cm in diameter; flowers 2 per cluster, ascending; lower inflorescence bracts 2–3, 2–4 cm long, membranous, tapering to an acute point on the lower half or two-thirds; bracteoles membranous, lanceolate or linear-lanceolate, 0.2–0.3 cm long, acute, spreading; pedicels 0.1–0.2 cm long, with the persistent part usually formed by the tapering base of the flower-tube. Flower white; perianth 2–3(–3.7) cm long. Fruit a globose berry. Fig. 2, p. 24.

var. **gracilis**

Leaves 25–75 cm long, to 0.9 cm thick. Inflorescence to 30 cm long. Flowers greenish grey, 25–30(–37) mm long.

UGANDA. Karamoja District: Warr, 5 Nov. 1931, *A.S. Thomas* 3182
KENYA. Northern Frontier District: Moyale, 29 July 1952, *Gillett* 13648!; Machakos District: Lukenya, 40 km from Nairobi on Nairobi–Mombasa rd., 24 Sep. 1978, *Gilbert* 5054!; Tana River District: Tana River National Primate Reserve main gate, 2.4 km N, 19 Mar. 1990, *Luke et al.* TPR 697!
TANZANIA. Musoma District: Serengeti National Park, S Musabi Plains, NW of Bingowe ridge, 27 Jan. 1974, *Kreulen* 309!; Handeni District: Kwa Mkono, 31 Aug. 1984, *Archbold* 3021!; Iringa District: Ruaha National Park, 10 Nov. 1970, *Richards* 26378!
DISTR. **U** 1; **K** 1, 4, 6, 7; **T** 1–3, 5–7; Ethiopia, Zambia, Malawi
HAB. Dry bushland and thicket, often on rocky outcrops in shallow soils, also in grassland and evergreen bushland; may form dense stands; 0–1800 m
USES. No data
CONSERVATION NOTES. Least concern

SYN. *Sansevieria caulescens* N.E.Br. in K.B. 1915: 200, Fig. 2a-b (1915); L.E. Newton in Ill. Handbook Succ. Plants 1: 263 (2001). Type: plant flowered at Kew in 1913, sent from Kenya, without locality, *Powell s.n.* (K!, holo.)

NOTE. This is the most variable species in growth habit of all our species. It ranges in size from very small to very large; leaves are smooth and banded in most Tanzanian specimens. The Kenyan material had many very large, slightly rough, unbanded individuals with many in between the two extremes.

var. **humbertiana** (*Guill.*) *Mbugua*, **stat. & comb. nov.** a var. gracili foliis minoribus plusminusve 33 cm longis basi crassioribus et divergentibus praecipue distinguenda; etiam inflorescentia brevior plusminusve 20 cm longa, flores breviores plusminusve 2 cm longi et griseo-brunnei (in var. *gracili* flores longiores viridi-grisei). Typus: Kenya, *Humbert* 233 no. 5 (P!, holo.)

Acaulescent herb, rhizomatous; stoloniferous. Leaves recurved, diverging, not at all distichous, ± 10, cylindrical and channelled above, up to 33 cm long and 1.3 cm thick, gradually and regularly tapering, smooth to rough, apex white, subulate, bottom of channel concave at the base, broadly triangular to nearly flat at the apex; margin `sharp', below the middle reddish or white, dorsal surface scarcely (hardly) grooved, dark green or marbled. Inflorescence greenish and white-spotted, almost 20 cm long, the lower one third bare; lower inflorescence bracts 3, 3–4 cm long, membranous, acute; flowers 2–3 per cluster; bracteoles membranous, white, small; pedicels 2 cm long, articulate at apex. Flower 2 cm long, buds spreading-recurved, white with greenish apex; perianth greyish brown; tube scarcely longer than the lobes; lobes linear; stamens equalling perianth; anthers yellowish-green; style exceeding stamens. Fruit a globose berry.

KENYA. Kitui District: Mutomo–Kibwezi road, 20 km S of Mutomo, 10 Apr. 1973, *Pfennig* 1108!; Teita District: Buchuma [Bachuma] Plain, Voi–Mackinnon Road, 15 Dec. 1978, *Bally* 12671! & Tsavo National Park, Mazinga Hill near Voi, 18 Aug. 1969, *Bally* 13416!

TANZANIA. Lushoto District: Mashewa, 26 Aug. 1954, *Faulkner* 1500! & Perani Forest, SE of Umba, 12 Aug. 1953, *Drummond & Hemsley* 3722!; Handeni District: Kwa Mkono, 28 Sep. 1966, *Archbold* 877!

DISTR. **K** 4, 7; **T** 3; Malawi

HAB. Dry bushland in rocky sites, thicket; 50–600 m

USES. No data

CONSERVATION NOTES. Least concern

SYN. *S. humbertii* Guill. in Bull. Mus. Hist. Nat. Paris ser. 2, 12: 353 (1940)

18. **Sansevieria ballyi** *L.E.Newton* in Brit. Cact. Succ. J. 22(1): 11, figs. 1–4 (2004). Type: Kenya, Kivuko Hill, *Newton* 5594 (K, holo.; EA, iso.)

Rosulate herb, with horizontal stolons. Leaves spreading, 6–10, laterally compressed-cylindric, 6–12 × 0.6–0.9 cm, apex a red-brown spine to 7 mm long, grooved on the face from base to 66–75 % of the length, with alternating transverse bands of dark and light green, margins of the groove with a red-brown line and a white edge, surface distinctly rough; scale leaves triangular, to 2 × 0.9 cm, acute. Inflorescence simple, erect, to 15.5 cm long; axis pale green, 3 mm in diameter, flowers 2 per cluster; bracts to 12 × 5 mm; bracteoles triangular, 3 × 3 mm; pedicel 1–2 mm long. Flowers with perianth tube 18–22 mm long, lobes 10–13 mm long; filaments to 12 mm, anthers 2–1.5 mm long; style white, exserted for up to 13 mm; stigma to 1 mm diameter.

KENYA. Teita District: Kivuko [Kivuto] hill, Apr. 1963, *Bally* 12681; idem, Oct. 1996, *Newton* 5594 & Manga hill, May 2003, *Newton* 5856

DISTR. **K** 7; not known elsewhere

HAB. Rock faces, in soil pockets among sedges, or below shrubs on hillsides; 150–1000 m

USES. No data

CONSERVATION NOTES. Data deficient

NOTE. Close to *S. gracilis* but differs in shorter stolons, shorter leaves with rough surface, distinctive colour banding and shorter inflorescence.

19. **Sansevieria suffruticosa** *N.E.Br.* in K.B. 1915: 202, fig. 3 (1915); Blundell, Wild Fl. E. Afr.: fig. 635 (1987); Mbugua in U.K.W.F. ed. 2: 312, t. 141 (1994); L.E. Newton in Ill. Handbook Succ. Plants 1: 271 (2001). Type: plant flowered at Kew in 1911 from material sent in 1903 from Kenya: without precise locality, *Evans* s.n. (K!, lecto., chosen here)

Semi-erect to procumbent herb; rhizomatous, with aerial rooting and branching; stem not more than 35 cm long, branching 2–7.5 cm above the ground, forming with the leaves clumps 60–80 cm high; aerial branches spreading or ascending, 7.5–25 cm long, 2–3.5 cm in diameter, rooting, covered on the lower part with very acute spine-tipped white-edged green scaly leaves 1.9–3.8 cm long, 4.5 cm wide, which gradually pass into the leaves. Leaves irregularly directed, sometimes ± distichous, both arrangements occasionally occuring on the same stem, ascending or spreading, (5–)7–18(–23) per plant, cylindric, 15–60 × 1.3–1.9 cm, rough throughout but smooth at the basal part and rough above, sheathing at the base and usually with a concave channel extending from sheath 1–2 cm way up the leaf, apical spine 0.3–0.6 cm long, gradually tapering to a very acute horny ending, brownish grey; dark green, faintly banded with pale green and marked with darker green in full sun but less marked in deep shade; abaxial surface continuous and interrupted with longitudinal lines, which with age or on shrivelled leaves become impressed, forming furrows. Inflorescence a spike-like green raceme, 30–38 cm long, axis 0.4–0.6 cm wide, thickly covered with minute whitish linear dots; flowers 2–5 per cluster; lower inflorescence bracts 2–4, membranous near base, 4.4–6.9 cm long on the upper inflorescence, acute; bracteoles ovate, 0.1–0.4 cm long, acute or acuminate, membranous; pedicels 0.3–0.4 cm long, jointed at or above the middle. Flowers whitish grey to greenish white at apices with reddish brown to dark brown outside; tube 1–2 cm long, slightly enlarged at the base, lobes (0.9–)1.2–1.4 cm long, recurved-spreading or with reflexed tips; anthers 2.5 cm long.

var. **suffruticosa**; L.E. Newton in Ill. Handbook Succ. Plants 1: 271 (2001)

Corolla tube ± 1 cm long.

KENYA. Naivasha District: Ilkek, near Gilgil, July 1969, *Tweedie* 3666!; Laikipia District: km 35, Nanyuki–Rumuruti [Thompsons Falls] road, 20 Apr. 1952, *Bally* 8178!; Nairobi District: Mbagathi gorge, 1 Oct. 1939, *Bally* 288!
TANZANIA. Musoma District: Moru Kopjes, 9 Feb. 1968, *Greenway, Kanuri & Brown* 13155!; & Iringa District: Ruaha National Park, at Nyamakuyu rapids, 6 Aug. 1970, *Bjørnstad* AB495!
DISTR. **K** 3, 4; **T** 1, 7; Ethiopia, Malawi
HAB. Dry or evergreen bushland, often on rock; 750–2200 m
USES. No data
CONSERVATION NOTES. Least concern

NOTE. This species is close in appearance to *S. phillipsiae*, sometimes even growing within the same populations in the wild e.g. at the Nairobi River Falls. The two are different in that *S. phillipsiae* has leaves that are smooth, unbanded, deep-dark green throughout and normally larger in size. The precise type locality may be inferred from where specimens with short flower tubes have been collected: around Nakuru. *S. suffruticosa* var. *suffruticosa* is found N of a line running south-east of the Mau escarpment, Aberdares and Mt Kenya.

var. **longituba** *Pfennig* in E.J. 102: 178 (1981); L.E. Newton in Ill. Handbook Succ. Plants 1: 271 (2001). Type: Kenya, Nairobi River Falls, *Pfennig* 1336 (EA!, holo.)

Corolla tube 2 cm long.

KENYA. Fort Hall District: Athi River, Aug. 1939, *Bally* 126!; Nairobi District: Nairobi River Falls, 14 Sep. 1977, *Pfennig* 1336!
DISTR. **K** 4; not known elsewhere
HAB. 'Scrub' on rocky ground near river-bed; 1250–1450 m
USES. No data
CONSERVATION NOTES. Data deficient

20. **Sansevieria bella** *L.E.Newton* in Cact. Succ. J. U.S.A. 72(4): 224 (2000) & in Ill. Handbook Succ. Plants 1: 262 (2001). Type: Kenya, Masai District: 24 km NW of Ewaso Ngiro junction, *Newton* 3945 (K!, holo; EA!, iso.)

Sprawling herb; stem trailing for up to 15 cm long with adventitious roots beneath, stem apex upturned and bearing leaves, with branch(es) just below apex to continue stem growth. Leaves up to 8, ± distichous, laterally compressed, to 70 × 3.5 cm, 2.5 cm thick, channelled on upper face, channel to 17 cm long, 10–15 mm deep near base, leaf surface rough, tapering to a red-brown spine to 5 mm long, sharp or blunt; lamina with dark and light green transverse bands and darker green narrow longitudinal lines, edges of channel a redbrown line with narrow colourless flange. Inflorescence simple, to 60 cm long, fertile for most of this, peduncle light green with slight bloom; up to 7 flowers per cluster; bracts triangular, to 9 × 4 cm; bracteoles triangular, 3 × 2 mm; pedicel to 5 mm long. Flowers white; perianth tube 10–15 mm long, lobes 13–18 mm long, with pale brown mid-stripe; filaments to 15 mm long; style exserted for 15–25 mm.

KENYA. Masai District: 4.4 km SW of Ewaso Ngiro junction, Aug. 1999, *Newton* 5689 & Oloyiangalani, Kojon'ga Gorge, Jan. 1996, *Newton & Foresti* 5559 & 30 km S of 'Corner Baridi', Sep. 1998, *Newton* 5658

Distr. **K** 6; not known elsewhere
Hab. Rocky bushland, degraded dry bushland; 1700–2050 m
Uses. No data
Conservation notes. Data deficient

21. **Sansevieria deserti** *N.E.Br.* in K.B. 1915: 208 (1915), as *deserti*. Type: Botswana, on the banks of the River Botletle in the Kalahari desert, *Lugard* 9 (K!, syn.); shore of Lake Ngami and the River Botletle, Apr. 1890, *Nicholls* s.n. (K!, syn.)

Rosulate herb; rhizome subterranean, 0.6–1.2 cm in diameter. Leaves erect, (4–)5–10, twisted-distichous, compressed-cylindric, 30–60 cm long, ± 4.5–6 cm broad at the sheathing base, gradually tapering to the horny white terete and sharp spine, apex 1–1.7 cm long, margins deep red-brown, less than 1 mm wide, acute, edged with white, thin membrane in some parts, very wide at base in old leaves in lower half, more or less obtuse and green towards the apex, the red-brown edge of the margin run like a collar around the base of the white apical point, slightly rough to very rough, channelled on the face and with ± 7–12 shallow grooves on the rounded back, thickness from the bottom of the channel to the back measured ± one third of the way up 0.8–1.5 cm; channel as wide as leaf near the base and gradually narrowing upwards, especially narrow on the inner (i.e. youngest) leaf; scaly leaves 1–5, 3–10 cm long, stiff, rounded apex, more or less deltoid, broad-based and clasping, red-brown margin with wide white-brown membrane. Inflorescence a terminal spike-like raceme ± 50 cm long; axis pale green, terete, 0.5–0.6 cm in diameter; flowers (3–)4–6 per cluster; lower inflorescence bracts 3, membranous, ± 5 cm long, clasping two-thirds of the circumference at the base of the peduncle; bracteoles 3–10 mm long, fleshy at the base, and thence produced into a membranous acuminate early withering scale; pedicels 5–8 mm long, jointed about the middle, the upper part early deciduous with flower. Flower buds cylindric, 2.1–2.4 cm long, with a lustrous appearance, drab yellow or white faintly flushed with pink or mauve and drab yellow or white faintly flushed with a mauve and the lightly clavate tip greyish-olive; tube ± 8–10 mm long, lobes linear, 12 mm long, obtuse, strongly revolute near the base; filaments yellowish, ± as long as the perianth, anthers 2.5–3 mm long; style whitish, ± 2 cm long. Fruit up to 2 per cluster, a globose berry 0.6–0.9 cm diameter. Fig. 3, p. 29.

Tanzania. Rungwe District: Kyimbila, 1915, *Stolz* s.n.!
Distr. **T** 7; Zambia, Mozambique, Zimbabwe, Botswana
Hab. no data for our area
Uses. No data
Conservation notes. Least concern

Note. This taxon was also seen at Brookside Nurseries, Kent, where it had a spiral-distichous leaf arrangement and a very abrupt twist. Newton in Ill. Handbook Succ. plants (2001) regards it as a synonym of *S. pearsonii* N.E. Brown.

22. **Sansevieria sordida** *N.E.Br.* in K.B. 1915: 214, fig. 8 (1915); L.E. Newton in Ill. Handbook Succ. Plants 1: 270 (2001). Type: Native country unknown, described from a living plant cultivated at Kew, flowered in March 1910, received from Mr *Bull* in Chelsea (K!, holo.)

Large rhizomatous herb; stem absent or up to 5 cm long, up to 2.6 cm in diameter, concealed in imbricating leaf-bases. Leaves distichous, 4–12, slightly spreading fanwise, straight to slightly recurved, slightly compressed-cylindric, 65–105 × 0.8–1.3 cm, 1.3–1.9(–2.2) cm thick from front to back, very rough, tapering to a sharp whitish or grey spine 0.8–1.3 cm long; leaf channelled down inner surface with acute dark brown edges but becomes shallow towards the base, membrane discontinous dark brown above, but whitish membrane towards the leaf base, with 11–15 or more grooves, dull bluish green with numerous darker longitudinal lines. Inflorescence a

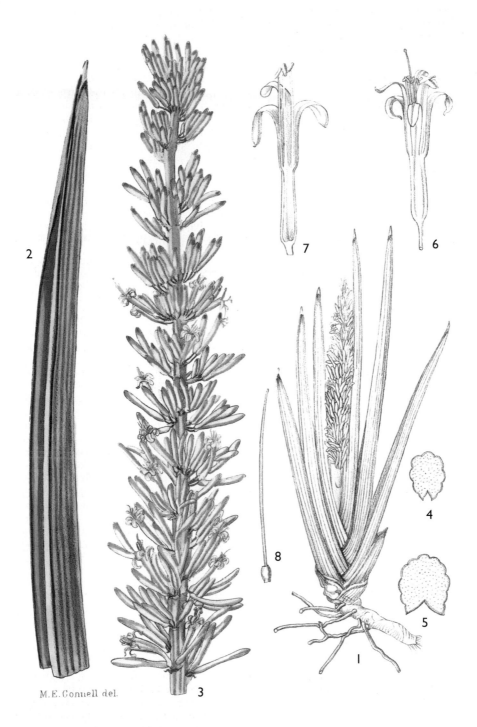

M.E. Connell del.

FIG. 3. *SANSEVIERIA DESERTI* — **1**, habit sketch; **2**, leaf, × ¹/₃; **3**, inflorescence, × ¹/₃; **4**, **5**, leaf cross section, × ¹/₃; **6**, flower × 2; ²/₃; **7**, flower with front petals removed, × 2; **8**, carpel, × 2. Drawn by M.E. Connell.

spike-like raceme 30–60 cm long, axis pale dull light green, 0.6 cm in diameter, minutely dotted with white; flowers 7–14 per cluster; lower inflorescence bracts 2–3, 1.9–3.8 cm long, tapering to a fine subulate point, produced on lower third of peduncle, distant, membranous; bracteoles subulate, 0.3–0.6 cm long; pedicels 0.4–0.5 cm long, persistent part 0.3–0.6 cm long, slender. Flower erect or ascending, pale dingy greenish in bud to purplish-blue on maturity; tube white to greenish, 0.7–1 cm long, lobes white on the inner face, green with minute dull purplish dots on the outside, linear, 1.2–1.6 cm long, obtuse, revolute, recurved in 3 lobes, but some in 5, becoming horizontal or distinctly drooping when expanded. Fruit a berry, greenish brown on ripening

KENYA. Teita District: between Voi and the Taita Hills, 1905, *Grenfell* 4!
DISTR. **K** 7; Zambia, South Africa
HAB. no data for our area
USES. No data
CONSERVATION NOTES. Least concern

NOTE. The number of flowers per cluster is very high, 7–14, with very prominent and persistent pedicels 0.5–0.6 cm long. This species resembles *S. volkensii* in its growth habit, but the leaves are much larger, smooth with very distinct longitudinal grooves. It is possible that this is a variant of *S. pearsonii* N.E.Br. that grows in Southern Africa, but the number of leaves and flowers are too high; *S. pearsonii* has a spiral-distichous arrangement with a slightly rough surface.

23. **Sansevieria patens** *N.E.Br.* in K.B. 1915: 210, fig. 5/a–b (1915); L.E. Newton in Ill. Handbook Succ. Plants 1: 268 (2001). Type: plant flowering at Kew in Apr. 1910, sent from "Tropical Africa", origin unknown, but probably Kenya fide N.E. Br. (K!, holo.)

Acaulescent herb with subterranean branching, rhizome 2–2.5 cm in diameter. Leaves spiral-distichous, 5–10, recurved-spreading, compressed-cylindric, 45–90 × 1.5–3 cm, 1.5–4 cm thick front to back, the outer gradually shorter, slightly rough, apex 0.6–1.3 cm long, hard and sharp but easily broken, white, with a brown band at base; channel from base to apex, several deep furrows running parallel to the channel all round the leaf, channel margins 2–15 cm from base, acute green whitish at base; leaf faintly banded when juvenile becoming bluish-green with age, numerous blackish-green longitudinal lines some to apex. Inflorescence a spike-like raceme of flower clusters, 45–60(–90) cm long, peduncle pale green (purplish blue when young), one-third of this length, with numerous greyish orbicular spots; flowers 2–3 per cluster; lower inflorescence bracts 2–3, 1.8–5 cm long, on basal third, acuminate; bracteoles lanceolate, 0.3–0.6 cm long, acute, membranous; pedicels 0.5–0.6 cm long, persistent part 0.3–0.4 cm long, jointed at the middle. Flowers grey-white, tube 0.9–1 cm cm long, 0.1 cm wide, slightly inflated at base, lobes linear, 1.2 mm long, 1 mm wide, revolute. Fruit a globose berry.

KENYA. plant flowering at Kew in Apr. 1910, sent from "Tropical Africa", origin unknown, but probably Kenya!
DISTR. Possibly **K** 7
HAB. no data
USES. No data
CONSERVATION NOTES. Data deficient

24. **Sansevieria kirkii** Baker in K.B. 1887: 3, 8, fig. 3 (1887) & in K.B. 1893: 186 (1893) & in Bot. Mag. 120, t. 7357, l.c. & F.T.A. 7: 334 (1898); Gérôme & Labroy in Bull. Mus. Hist. Nat. Paris 1903: 170, 173, fig. 7 (1903); De Wildem., Not. Pl. Utiles Congo: 622, 624, 625, fig. 7 and pp. 632, 635; N.E. Brown in K.B. 1915: 254, fig. 23 (1915); T.T.C.L.: 22 (1949); U.O.P.Z.: 440 (1949); L.E. Newton in Ill. Handbook Succ. Plants 1: 267 (2001). Type: plant flowered at Kew in June 1893 from material sent from Tanzania, Pangani District: Pangani [Pangane], *Kirk* s.n. (K!, lecto., selected here)

Fig. 4. *SANSEVIERIA KIRKII* — **1**, leaf, × ¹/₃; **2–5**, cross-sections of leaf, 2 from near base and others from gradually higher up at respectively 40, 80 and 120 cm above base, × ²/₃; **6**, inflorescence, × ¹/₃; **7**, flower, × ²/₃. All from type specimen. Drawn by Juliet Williamson.

Acaulescent herb; rhizome subterranean, crowded with scaly leaves. Leaves 1–3, erect or ascending-spreading to upper part recurved or even drooping, elongate-lanceolate or broadly strap-shaped, 45–180(–270) × 5.8–8.8 cm, 0.8–1.4 cm thick at the middle, very rigid to stiffly coriaceous towards acute apex, tipped with a horny spine 0.8–1.3 cm long, very wavy along the sides, margins reddish brown, hardened, edged with white in juvenile, at length often breaking up into thread-like fibres, smooth, in juveniles spreading to recurved-spreading, shortly petiolate, concave to flattish, lanceolate; midrib broad, rounded on the back; petiole 2.5–3.1 cm broad, 3.8 cm thick; basal scaly leaves membranous, distinctly parallel veined. Inflorescence capitate-corymbose, (25–)37–60 cm long, axis dull purplish brown, 1.3–1.9 cm in diameter, thickly speckled with pale green or dull whitish; inflorescence axis 3.8–6.3 cm long, capitate, dense; lower inflorescence bracts distant, green, ovate or lanceolate, (4–)5–6, 3.5–12.5 × 3.8 cm, the upper spreading from stalk, speckled with fuscous or dull purplish brown at the tips; lower bracts lanceolate, oblanceolate, ovate or oblong-ovate, 2.5–3.8 × 0.8–1.9 cm, others gradually smaller, acute or subobtuse, green, suffused with dull purplish brown and dotted with paler brown; pedicels 6–10 mm long, 1.5–2 mm in diameter, not jointed. Flower erect or ascending, green on the back and tips, the rest brownish-pink, 12–13(–14.5) cm long; tube purplish or dull pink, ± 11–12.5 cm long, lobes linear, 3–4.5 cm long, obtuse, spreading, with revolute tips; filaments 17–17.5 cm long, anthers 4–5 mm long; style 15.5–16.5 cm long, exserted beyond the lobes. Fruit a globose berry. Fig. 4, p. 31.

var. **kirkii**; L.E. Newton in Ill. Handbook Succ. Plants 1: 267 (2001)

Leaves greyish green, light grass-green and greyish green beneath, as if densely dusted with green and whitish or pale-green, slightly shining above, opaque beneath; 3–4 longitudinal dark green lines, 3–9 on under-surface. Inflorescence bracts ovate or oblong-ovate; lower inflorescence bracts ovate, 5–7.5 cm long.

KENYA. Kwale District: Kaya Fungo, June 1994, *Luke* 4012 & June 1994, *Luke* 4013
TANZANIA. Lushoto District: Amani–Muhesha road, 4 km E of Sigi railway station, 27 July 1953, *Drummond & Hemsley* 3504!; Ufipa District: Sumbawanga, Kasanga, 25 Nov. 1959, *Richards* 11839!; Iringa District: Livingstone Mountains, lower valley of Ngolo River, 3.4 km above Lupingu, 1 Feb. 1991, *Gereau & Kayombo* 3825!; Pemba: Ngezi Forest, 1 Oct. 1902, *Lyne* 105!
DISTR. **K** 7; **T** 3, 4; **Z**, **P**; Gabon, Zambia, Malawi, Mozambique
HAB. Forest, thicket, riverine fringe, often on thin soil among rocks or on coral; 0–900 m
USES. Sir John Kirk said in 1881 it "... yields a most excellent fibre"
CONSERVATION NOTES. Least concern

SYN. *S. aubrytiana* Carr. in Rev. Hort. 1861: 448 (1861), *nomen nudum*
　　　S. aubrytiana Gérôme & Labroy in Bull. Mus. Hist. Nat. Paris 9: 169, 173, fig. 9 (1903). Type: Gabon, *Aubry-Lecomte* s.n. (P, holo.)

NOTE. Cultivated specimens tend to have longer and more pronounced petioles than specimens from the wild.

var. **pulchra** *N.E.Br.* in K.B. 1915: 256 (1915); T.T.C.L.: 22 (1949); U.O.P.Z.: 236, fig. (1949); L.E. Newton in Ill. Handbook Succ. Plants 1: 267 (2001). Type: plant flowering at Kew in Sep. 1912, collected in Tanzania, Zanzibar by *Last* in 1903 (K!, holo.)

Leaves (especially the younger) handsomely and conspicuously marked with whitish green or somewhat buff-coloured or sometimes almost reddish spots or irregular bands on both sides and with a white membranous edge to the red-brown wavy margin; firm. Inflorescence bracts lanceolate-oblanceolate; lower inflorescence bracts lanceolate, 3.5–12.5 cm long.
NB: Colour disappears on herbarium specimens but they are dark green or blackish and narrower, underside of leaf pale pink with dark-green bands.

TANZANIA. Pemba, 1901, *Vaughan* 719!; Zanzibar, Chwaka, 17 May 1960, *Faulkner* 2558! & Chwaka, 18 Sep. 1974, *Raadts* s.n.!
DISTR. **Z**, **P**; Congo-Kinshasa, Zambia

HAB. On coral near the sea; near sea level
USES. No data
CONSERVATION NOTES. Least concern

SYN. *S. longiflora* sensu Gérôme & Labroy in Bull Mus. Hist. Nat. Paris 1903: 173, fig. 8 (1903); De Wildem., Not. Pl. Utiles Congo: 624, 630, fig. 8, *non* Sims

NOTE. A specimen of this taxon was grown at Kew Gardens (*Brown* 203/1904) and flowered 25.x.1910, mentioned under the name *S. longiflora* by N.E. Brown. Provenance is unclear.

25. **Sansevieria pfennigii** *Mbugua* **sp. nov.** *S. fischeri* arcte affinis sed foliis 2 plerumque minoribus 0.4–0.5(–0.6) cm crassis (39–)44–64(–76) cm longis angustissimis sulcis numerosis obsitis sed carina carentibus, rhizomate parvo plusminus 1.5 cm crasso, foliis squamosis basalibus 2–3 acutis membranaceis 2.5–4 cm longis, inflorescentia corymboso-capitata pedunculo vaginis obsito, bracteis longissimis 1.3–1.7 cm longis flores amplectantibus, floribus in axillis bractearum et in internodiis inter bracteas exorientibus distincta. Typus: Tanzania, Lindi District: Lake Lutamba, *Schlieben* 5917 (MO!, holo., K!, iso.)

Herb with subterranean rhizome 1.5–1.8 cm in diameter. Leaves erect, 1–2 per plant, cylindrical from base to apex, 44–64(–76) × 0.4–0.6 cm, slightly recurved at base then straight and inward recurved just before apex, smooth to slightly rough towards apex, spine 0.3–0.4 cm long, brownish red at base, grey-greenish towards the tip, sharp but hardly piercing, horny; no channel, grooves numerous formed on drying; basal scaly leaves 2–3, 2.4–4.3 × 1.2–1.4 cm, acute, membranous. Inflorescence terminal, corymbose-capitate, 11–13.5 cm long; axis 0.4–0.6 cm in diameter; one flower per bract, ensheathed in the bracts, no evident clusters, lower flowers seem to be sterile; lower inflorescence bracts 5–7, 0.6–2.5 cm apart, 0.9–1.3 cm long, membranous, acute; bracteoles densely together, 1.2–1.6 × 0.4–0.6 cm, acute, membranous. Flower perianth white-pink or greyish green, 3.5–4.2 cm long, 0.2 cm wide; stamens at level of lobes, anthers 0.4 cm long; style exserted beyond the lobes. Fruit a globose berry, blackish when dry.

TANZANIA. Pangani District: Pangani, Nov. 1889, *Stuhlmann* 1126a!; Lindi District: 10 km W of Lindi, near Lake Lutamba, 28 Jan. 1935, *Schlieben* 5917! & Mto Nyangi Hippo Pools, Nov. 2003, *Luke & Kibure* 9773!
DISTR. T 3, 8; not known elsewhere
HAB. Near lake-shore, swamp and pools; sometimes in large groups; 50–250 m
USES. No data
CONSERVATION NOTES. Data deficient

NOTE. Named after Dr. H. Pfennig who devoted a great part of his life to the study and collection of sansevierias.
    Seen as part of *S. sulcata* Baker by Newton in Ill. Handbook Succ. Plants 1: 271 (2001) but that has a channel running from top to bottom i.e. slightly crescentic and also has a longer peduncle with flowers further apart from each other.

26. **Sansevieria stuckyi** *God.-Leb.*, Sansev. Gigant. Afr. Orient.: 13, 17, 33, figs (1861); N.E. Brown in K.B. 1915: 219, fig. 10 (1915); Gérôme and Labray in Bull. Mus. Hist. Nat. Paris 1903: 64 (1903); L.E. Newton in Ill. Handb. Succ. Pl. Vol. 1: 270, fig. 29e (2001). Type: Mozambique, Boror, *Stucky* s.n. (K!, holo.)

Acaulescent herb; rhizome subterranean. Leaves stiffly erect, 1–2(–3), cylindrical, 120–270 cm long, (2.9–)3.8–6.3 cm thick at base, slightly rough, apical spine pale brown, 0.4 cm long, horny, hardly piercing; channel with acute green edges from base to apex, 0.4–0.8 cm deep, 6–3.3 cm broad at base but 1–1.7 cm broad near the apex; 6–20 longitudinal lines, dull green, faintly banded, slightly glaucous. Inflorescence emerging from between the leaves, erect, crowded, corymbose-capitate, 5–9 cm long, axis 0.5–0.7 cm wide; flowers 1–3, erect, hardly in clusters; lower inflorescence bracts 4, purple, broadly lanceolate, 3.5–6 cm long; bracteoles greyish green, 0.7–1.3 cm long, membranous; pedicels 1.1–1.5 cm, loosely jointed at base. Flower perianth 7.5–12.5 cm long, 0.3 cm wide, thickened at base to 0.5 cm, the lower flowers (outmost) are shorter, purple speckled with green; tube 8–9.5 cm long, lobes grey-white, linear, 3–3.5 cm long, rolled; stamens 8–8.5 cm long, anthers 0.4 cm long; style 8–15 cm long. Fruit a globose berry.

KENYA. Masai District: around Laitokitok [Loitokitok], 8 Mar. 1982, *Pfennig* 90!
DISTR. **K** 6; Mozambique
HAB. In the undergrowth on red and sandy soils in the lowlands; no altitude given
USES. No data
CONSERVATION NOTES. Least concern?

NOTE. A species that differs from *S. fischeri* by having more than one leaf per plant with a wide continuous channel from base to apex. The inflorescence arises between the leaves and is usually more racemose-capitate than in *S. fischeri*, which forms a wide crown on a rather long peduncle.

27. **Sansevieria fischeri** (*Baker*) *Marais* in K.B. 41(1): 5 (1986); Thulin, Flora Som. 4: 29 (1995); Demel in Fl. Eth. & Eritr. 6: 80 (1997); L.E. Newton in Ill. Handbook Succ. Plants 1: 265 (2001). Type: Tanzania [Ostafrika], *Fischer* 9 (B!, holo.; K!, fragment)

Acaulescent herb; rhizome subterranean, 5–9 cm long, 2–3 cm in diameter. Leaves solitary, erect, rigid, cylindric, (0.8–)1.5–2(–3) m long, 2–3 cm thick at the middle, slightly rough, slightly tapering upwards to apex, narrowing to a stout acute whitish hardly prickly point; juvenile with a concave channel 0.4–0.6 cm wide, 0.2–0.4 cm deep on upper surface and almost closed at leaf apex; 4–6 longitudinal impressed lines on the sides and back which deepen into furrows with age, dull green or bluish green, often with a brownish tint and entirely greyish green when mature, slightly subglaucous especially when young, and then marked with numerous closely placed transverse pale green bands which disappear with age; margin in juveniles brownish red to grey-green with a whitish grey membrane on the outer side, the brownish grey margin disappear with age; basal scaly leaves flat, very reduced, ensheathing the solitary cylindrical leaf, 8–10, 6.5–10(–13) × 2.5–3.5 cm, membranous, subulate tip less than 0.1 cm long, very evident parallel veins, translucent membranous grey-white margins (brownish white on live material); 7 back grooves, 3–4 lines in the channel. Inflorescence lateral, emerges beside the leaf, crowded corymbose-capitate, 8–10 cm long, open crown diameter up to 12–19 cm; flowers appear solitary above, erect; bracteoles 0.4–0.8 cm long, at base of cluster or each flower; base of cluster 0.3 cm thick; pedicels 0.6–0.9 cm long, very loosely jointed at base. Flower white sometimes tinged with violet, perianth 2.8–3.5 cm long, 0.2–0.3 cm wide; tube 1.2–1.6 cm long, lobes 0.8–0.9(–1.3) cm long; stamens 1.2–1.4 cm, anthers 0.3 cm; style 2.4–2.6 cm long. Fruit a globose berry, 1.3–1.5 cm diameter; seeds 1 × 0.7 cm.

KENYA. Teita District: Tsavo National Park East, Voi Gate–Sala Gate via Aruba ± 20 km, 17 Jan. 1967, *Greenway & Kanuri* 13056! & between Voi and Taita, 2 km E of Maktau, 16 May 1971, *Pfennig* 1019! & Voi, 6 Mar. 1953, *Sheldrick in Bally* 8592!
TANZANIA. "probably not far from Dar es Salaam (obtained through friends)", 13 Feb. 1994, *Mbugua* TZ 20!
DISTR. **K** 7; **T** 6; Ethiopia, Somalia
HAB. Dry bushland; may form dense stands; 250–550 m
USES. Much liked by elephants (fide Greenway)
CONSERVATION NOTES. Least concern

SYN. *Buphane fischeri* Baker in F.T.A. 7: 577 (1898)
    Sansevieria singularis N.E.Br. in K.B. 1911: 97 (1911) & in K.B. 1915: 222 (1915). Type: Kenya, Teita District: Voi, *Powell* 2 (K!, lecto., chosen here)

NOTE. The growth habit of this species is quite diagnostic in that it never produces more than one leaf at any one point. The banding, however, is very variable depending on whether it is in the open or in the undergrowth; the undergrowth specimens tend to be taller, narrower and dark green with distinct light grey bands whereas those in the open are shorter, thicker and light green to grey-green without any visible banding. It was observed, however, that some specimens remained banded and deep green even after being grown in full light for over one year. This phenomenon needs further investigation with more material from the wild. The inflorescence is described here for the first time.

The syntype collected by *Tompson* and said to be from Uganda is actually from Kenya.

28. **Sansevieria pinguicula** *Bally* in Candollea 19: 145 (1964); L.E. Newton in Ill. Handbook Succ. Plants 1: 269 (2001). Type: Kenya, Northern Frontier District: Tana River region, near Bura, Jan. 1943, *Bally* 4275 (K!, holo.; EA!, iso.)

Erect short-stemmed herb 30 cm high. Leaves rosulate, 5–7, in cross-section semicircular, 12–30 cm long, 2.8–3.5 cm wide (thickness just below the middle), thickly fleshy, uniformly green, the side facing inwards concave-angular with horny edges, the outer side convex and 2–7-dented, tapering towards the cuspidate, horny, very sharp spine, upper surface deeply channelled, margins of the channel sharp, brown, horny, underside with 2–7 more or less well defined longitudinal grooves, rounded. Inflorescence a subterminal erect panicle 15–32 cm long, axis 5–6 mm diameter at the base, equalling or hardly exceeding the leaves, gradually tapering towards the apex, branched in the upper half; flowers 4–6 per cluster; lower inflorescence bracts 1–2, the upper ones subtending the branchlets; branchlets 6–9, the lowest to 5 cm long, the upper branchlets increasingly shorter, horizontally spreading; bracteoles broadly triangular, acute, membranous; pedicel 1.5–2 mm long. Flower perianth tube cylindrical, 4–5 mm long with a slightly inflated base, lobes 3–4 mm long, 0.7–1 mm wide; stamens equalling the perianth, filaments free from the base of the lobes, anthers 1.6–2 mm long; style slightly exceeding the anthers. Fruit a globose berry, very few reaching maturity.

KENYA. Northern Frontier District: Tana River region, near Bura, Jan. 1943, *Bally* 4275! & 24 km on the Garissa–Liboi road from junction with Mado Gashi road, 28 Jan. 1972, *Smith* & *Bally* 14976!; Tana River District: Bura, 8 May 1985, *Bally* 9386!
DISTR. **K** 1, 7; only known from Bura area
HAB. Open dry bushland; 50–250 m
USES. No data
CONSERVATION NOTES. Least concern

NOTE. This is called the walking *Sansevieria* because of producing aerial shoots followed by rooting. It is a very stout plant found in one of the most arid areas in Kenya. To date the only wild material has been obtained from Bura near Garissa. It is yet to be established whether it occurs elsewhere.

29. **Sansevieria robusta** *N.E.Br.* in K.B. 1915: 207 (1915); T.T.C.L.: 22 (1949); Mbugua in Cactus Succ. J. U.S.A. 66: 87 (1994) & in U.K.W.F. ed. 2: 312, t. 141 (1994); L.E. Newton in Ill. Handbook Succ. Plants 1: 269 (2001). Type: Kenya, Teita District: between Voi and Taita hills, *Grenfell* 13 (K!, lecto., chosen here)

Shrubby herb; rhizomes underground, yellowish white, quite stout and tough, growing upright ± 20–30 cm away from the parent plant; stem erect, light green-yellow, ± 30–50(–60) cm long, 3–5(–7) cm in diameter, with distinct nodes 2–3 cm apart, usually fully covered by leaf bases especially in young plants. Leaves distichous, erect, 6–14(–16) per plant, up to 20 in juvenile stages, upper side V-shaped, 1–1.5(–2.5) m long, width up to 3.5(–5) cm, smooth, apex with a sharp hard spine, margin dark brownish red, up to 2 mm wide, hardened, dark green with very light grey-white bloom on young plants; juveniles spirally arranged. Inflorescence a panicle with numerous branches, 0.8–1.4 m long, 0.5–0.7 m in diameter, stalks greenish grey, to 1.2–1.5 cm in diameter; flowers (4–)5–6(–7) per cluster, distance between clusters 1–2.5 cm; bracteoles to 2 mm long, acute, membranous, quickly deciduous; pedicels 0.3–0.4 cm long. Flower perianth 0.4–0.6 cm long; tube 0.3 cm long, lobes greenish grey to greenish brown on the outside, greenish yellow on the inside, 0.3 cm long, producing copious nectar. Fruit an orange-yellow berry, with little flesh.

KENYA. Naivasha District: Kedong escarpment, near Kedong River, 10–20 km S of Nairobi–Naivasha road, Jan. 1939, *van Someren* 8505!; Teita District: between Voi and Taita Hills, 31 Mar. 1906, *Grenfell* 8! & Voi, Mar. 1906, *Powell* 1!
TANZANIA. Pare District: Kisuani–Gonja, 4 Feb. 1930, *Greenway* 2148! & Kifaru Hill, 16 Jan. 1945, *Bally* 4255!
DISTR. **K** 3, 7; **T** 3, 8; Congo-Kinshasa
HAB. Rocky *Acacia-Commiphora* bushland, bushed grassland; 550–1950 m
USES. Fruit eaten by birds and other animals
CONSERVATION NOTES. Least concern

SYN. *Sansevieria perrotii* O.Warb., Tropenfl. 5, 4: 191, fig. (1901); N.E. Brown in K.B. 1915: 206 (1915); De Wild., Not. Pl. Utiles Congo: 626, 633 (1905); T.T.C.L.: 22 (1949); L.E. Newton in Ill. Handbook Succ. Plants 1: 269 (2001). Type: Tanzania, Lindi District, *Perrot* s.n. (not found)

NOTE. This species yields an abundance of fibres of good quality (Brown, 1915). The flowers are highly fugaceous even before opening, sometimes over 90% drop off. Each of the branches has a large bract subtending it. This was observed only on live specimens otherwise only the tiny base remains are seen on dry specimens, if at all. This species may be distinguished from *S. ehrenbergii* by having more open, completely smooth, brown greenish yellow leaves in dry specimens. The leaves have a transluscent and peeling epidermal surface while the fibres are not as rigid as in *S. ehrenbergii*.

30. **Sansevieria ehrenbergii** *Baker* in J.L.S. 14: 549 (1875); N.E. Brown in K.B. 1915: 207 (1915); Thulin, Flora Som. 4: 28, fig. 18 (1995); Demel in Fl. Eth. & Eritr. 6: 80 (1997); L.E. Newton in Ill. Handbook Succ. Plants 1: 264 (2001). Type: Sudan, between Atbara and the Red Sea, *Schweinfurth* s.n. (B, holo.)

Fruticose perennial; rhizomes subterranean, 2.2–2.6(–3.6) cm thick; stem (15–)20–70(–90) cm long, concealed by leaf bases, erect, terete, with 1.5–3.6 cm long internodes, 2.8–3.1 cm wide, tough fibrous, yellowish grey on outer surface. Leaves crowded, distichous, 5–9(–13), laterally compressed, with flattened sides, rounded on the back, (0.5–)0.8–3(–3.3) m long, 2.5–3.5 cm thick from side to side, 3–4.5 cm thick from front to back, glaucous especially in juvenile and young leaves, erect or less spreading (more erect than in *S. robusta*), smooth to rough surface, tapering upwards; rather abruptly ending in a spine 0.8–1 cm long, stout, horny, hardly piercing, basal brown all round; with a triangular channel as broad as the leaf all down the face and 5–12 shallow grooves, very faint or impressed lines down the sides and back; dark green to bluish green, the longitudinal lines blackish-green, no transverse bands; channel margins 1–2 mm wide, acute, reddish brown with small white membranous edges, usually spreading to a slightly wider breadth than the rest of the basal part. Inflorescence paniculately branched in the upper three-quarters of its length, (0.9–)2–3 m long, inflorescence stalk 0.6–1.1 cm thick, branches up to 3–4 from the same point i.e. a central major, 2 large and a smaller one, 5–7 cm apart; inflorescence stalk wavy, ascending, the lower branched, the upper simple; long bract 1.1–1.5 cm long, subtends each branch; flowers 4–5(–7) per cluster, clusters numerous, 1.4–1.8 cm apart, arranged in a spiral; bracteoles 0.1–0.2 cm long; pedicels 0.3–0.4 cm long, jointed above the middle. Flowers greyish purple, 1.1–1.3 cm long, tube 0.5–0.6 cm long, lobes linear, (0.3–)0.5–0.9 cm long, obtuse; style 1.4 cm long, thin; filament ± 1 long, anthers 0.2 cm long, yellowish. Fruit a berry, yellowish green when ripe. Fig. 5, p. 37.

FIG. 5. *SANSEVIERIA EHRENBERGII* — **1**, habit sketch; **2**, leaf cross section, × ²/₃; **3**, inflorescence, upper third, × ²/₃; **4**, fruits, × 1. All from *Tweedie* 3757. Drawn by Juliet Williamson.

Uganda. Karamoja District: near Nabilatuk, Feb. 1936, *Eggeling* 2920! & Upe County, Amudat, Kamjagareng River, 21 Feb. 1953, *Dawkins* 792! & Karamoja, Nov. 1945, *Thomas* 4428!
Kenya. Northern Frontier District: Moyale, 20 Aug. 1952, *Gillett* 13732! & 2 km south of Lokori Mission, 18 Aug. 1969, *Mwangangi* 1463!; Teita District: Voi, 1 Feb. 1953, *Bally* 8651!
Tanzania. Masai District: Anata Sikea, 2 Jan. 1971, *Richards & Arasululu* 26446!; Tanga District: Chakachani, 30 Jan. 1942, *Greenway* 6435!; Uluguru District: Morogoro, 18 km NE of Kingolwira station, 1 Mar. 1957, *Welch* 373!
Distr. **U** 1; **K** 1, 2, 7; **T** 2, 3, 6; Sudan, Ethiopia, Djibouti, Somalia; also in Yemen & Oman
Hab. Dry wooded grassland, dry bushland, thickets, often on rock or termite hills, also in coastal bushland or drier types of evergreen bushland; may be locally common, can withstand heavy grazing pressure; 0–1900 m
Uses. Used for fibre; Turkana use the thread for beadwork; the juice is used for minor wounds; roots are eaten by porcupine and gazelle
Conservation notes. Least concern

Syn. *Sanseverinia rorida* Lanza in Boll. Ort. Bot. Palermo 9: 208 (1910). Type: Somalia, near Mogadishu, *Macaluso* s.n. (PAL., holo.)
   Sansevieria rorida (Lanza) N.E.Brown in K.B. 1915: 205 (1915)
   Acyntha rorida (Lanza) Chiov., Result. Sci. Miss. Stef.-Paol, Coll. Bot.: 170 (1916)

Note. Clusters more prominent than in *S. robusta* and closer together. This species has leaves more upright than in *S. robusta*.

31. **Sansevieria bagamoyensis** *N.E.Br.* in K.B. 1913: 306 (1913) & in K.B. 1915: 197 (1915); T.T.C.L.: 21 (1949); L.E. Newton in Ill. Handbook Succ. Plants 1: 262 (2001). Type: Tanzania, Bagamoyo District: near Bagamoyo, *Sacleux* 672 (P!, holo., K, drawing!)

Erect frutescent herb, stem ± 1.5–2.5 m high, 30–40 cm thick, rhizome subterranean; rhizome (0.8–)1.5–2(–2.5) cm long, (1.3–)3–4 cm wide, terete, very tough, noded, completely sheathed and only a very short basal part exposed. Leaves distichous to spiral, 8–10(–12), linear-lanceolate, (0.5–)0.8–1(–1.5) m long, 2–2.6(–3) cm wide (along the V-shape), canaliculate down the face, very narrow, less than 0.1 cm thick, sheathed, recurved-spreading from the base, smooth; margin red-brown (brown-orange on drying), edged with a white membrane becoming wider towards leaf base; leaf apex 0.4–0.6 cm long, acute with a horny spine-like brownish yellow point; numerous veins evident on abaxial surface. Inflorescence a terminal panicle 40–55(–80) cm long, stalk 2.5–3(–5) cm diameter, with ± 13(–15) branches with ± 15–20 flower clusters per branch, laxly branched from 3.4 cm above the base; lower branches 14–20 cm long and 2.4–3.4 cm apart, upper 6–10 cm long and ± 1.8–2.4 cm apart, simple; bracts at branch bases, lanceolate,1.7–2.3 cm long, becoming smaller up the inflorescence, acute, membraneous towards inflorescence apex; apex 0.2 cm long, white, horny tip; bracts subtending clusters are smaller, 0.2–0.3 cm long; flowers 2–4 per cluster; pedicel 0.2–0.3 cm long and jointed at the middle, very thin and weak. Flowers grey-brown; whole flower 1.2–1.5 cm long; tube 0.5–0.7 cm long, 1.1 cm wide; lobes 0.5–0.6 cm long; style 1.3–1.6 cm long, filiform, exserted beyond the lobes; stigma globose; filament 0.6–0.7 cm long; stamen 0.3 cm long. Fruit a globose berry.

Kenya. Kwale/Kilifi District: Maji Ya Chumvi, 1929, *Graham* 1808!; Kwale District: Mwachi Forest, near Mwachi River, 17 Feb. 1977, *Faden et al.* 77/460! & Lunga Lunga at Kenya–Tanzania boundary, 10 April 1992, *Mbugua* 462!
Tanzania. Lushoto District: Mashewa–Magoma road, 8 km SW of Mashewa, 8 July 1953, *Drummond & Hemsley* 3228!; Handeni District: near Kideleko, 11 Oct. 1976, *Bally* 12309!; Uzaramo District: Msasani, 18 Feb. 1939, *Vaughan* 2755!
Distr. **K** 7; **T** 3, 6; not known elsewhere
Hab. Open forest, coastal bushland, thicket; 0–400 m
Uses. Good fibre
Conservation notes. Least concern

NOTE. This species may be easily distinguished from *S. arborescens* by its leaf, inflorescence and pedicel lengths being greater. The leaves are more outspread to recurved and light grey-green. It may be distinguished from *S. robusta* by having more outspreading-recurved and spirally arranged leaves with light greenish brown, slightly rough surfaces.

32. **Sansevieria powellii** *N.E.Br.* in K.B. 1915: 198, fig. 1 (1915); Thulin, Flora Som. 4: 28 (1995); L.E. Newton in Ill. Handbook Succ. Plants 1: 269 (2001). Type: Kenya, Kwale District: Mackinnon Road, cultivated at Kew, *Powell* 5 (K!, holo.)

Erect frutescent herb, rhizome subterranean, 1.7–2.4 cm thick; stem erect, terete, 100–200 cm long, (2.5–)3–5(–6) cm in diameter, stout, leafy especially the top three quarters of the plant, in some juvenile populations half of the plant is leafy, marked with ring-like scars for ± 15–30(–50) cm from the base; internodes 2–3 cm long, light brown-grey. Leaves distichous to spiral-distichous, 25–35(–40), very stiff, spreading to slightly recurved from half length, rounded to keeled on the back, 35–55(–60) cm long, 4–5(–6) cm wide, 1–2 cm thick, smooth to slightly rough, base sheathed, gradually tapering to a apical spine, horny pale brown with dark brown-red base, very sharp; deep channel from base to apex and as broad as the leaf; margin narrow red-brown edged with grey-white membrane widest towards leaf base, faintly glaucous; very faint markings but not banded, grass-green to greenish grey (when dehydrated) dark bluish green with age, numerous veins on abaxial surface. Inflorescence a panicle with 20–25(–30) branches, peduncle (45–)70–100(–120) cm long, 0.5–1 cm wide; branches up to 60 cm long near base of inflorescence and much shorter above, ascending-spreading, lower branches bear 3–10 branchlets; main stalk rough basally; bracts 0.1–0.2 cm long, fleshy, convex with a membranous acute apex; nectar exudes from bract bases; lower inflorescence bracts lanceolate, 0.5–1.7 cm long, acute, horny apex, 0.2 cm long; lower inflorescence bracteoles are thick, brownish grey but towards inflorescence apex are membranous and greyish-white, margins greyish-white also; flowers 4–6; pedicels 0.1–0.2 cm long, jointed at or slightly below the middle. Flowers dingy greenish white with dull brownish purple slender lines outside, perianth 1–1.6 cm long; tube 0.6 cm long, 0.2 cm wide above the slightly swollen base; lobes linear, (0.6–)0.9–1 cm long, obtuse; anthers 0.2 cm long, pale greenish yellow; style 1–1.1 cm long; stigma globose; ovary 0.2 cm wide. Fruit 1–2 per pedicel with the double being joined to each other at base, a globose orange berry 0.5–0.7 cm diameter.

KENYA. Tana River District: Galole–Malindi road turn-off for Garsen, 18 June 1972, *Gillett* 19970! & Tana River National Primate Reserve, 20 Mar. 1990, *Luke et al.* TPR 754!; Kwale/Kilifi District: Maji ya Chumvi, 17 Oct. 1976, *Bally* 12310!
TANZANIA. Musoma District: Serengeti National Park, Banagi, Mgungu river, 22 Feb. 1967, *Braun* 157!
DISTR. **K** 1, 7; **T** 1; Sudan, Somalia, Ethiopia
HAB. Dry bushland, scattered tree grassland; 0–550(–1350) m
USES. Making ropes by Amer people of Sudan
CONSERVATION NOTES. Least concern

SYN. *Acyntha powellii* (N.E.Br.) Chiov., Fl. Somala 2: 422 (1932)

NOTE. The protologue by Brown reads: "Although near allied to *S. caulescens* N.E.Br. and resembling it in appearance, it is easily distinguished by the channel of the leaves being as broad as the full width of the leaf and by its paniculate (not spike-like) inflorescence". This is not correct in my view as *S. caulescens*, for example, is not quite frutescent nor does it ever produce distichous fibrous light grey-green leaves. *S. powellii* is also a larger plant in all aspects.
　　It has been hypothesized by Chahinian and Pfennig (pers. comm.) that *S. powellii* is a hybrid between *S. robusta* and *S. arborescens*. This seems plausible, but it is yet to be proved experimentally since no *S. arborescens* were observed in most of the areas where *S. powellii* was growing. It might be that offspring colonised totally new areas i.e. intermediate altitudes. *S. robusta* for example was mainly found at mid to high altitudes, *S. arborescens* was predominantly ator near sea level, while *S. powellii* occupies the areas in between.

This species may easily be distinguished from *S. bagamoyensis* by its shorter and wider leaves with very sharp spines. The stem is generally longer and thicker with a deep brownish-orange colour on the internodes. Leaves on herbarium sheets are greyish brown, very stiff, more fibrous and with deep longitudinal surface folds. Specimens from Ethiopia tend to have very many double and triple fruits. This may be an indicator that there may be a more effective pollinator agent(s) in that area compared to other regions.

33. **Sansevieria arborescens** *Gérôme & Labroy* in Bull. Mus. Hist. Nat. Paris 9: 170, 173, fig. 20 (1903); N.E. Brown in K.B. 1915: 198 (1915); L.E. Newton in Ill. Handbook Succ. Plants 1: 262 (2001). Type: Tanzania, Zanzibar, *Sacleux* s.n. (P, holo.)

Frutescent, grows in thick stands, rhizomes subterranean; stem upright, terete, 1–1.5 m high, 2–2.5 cm in diameter, no aerial branching; node (1.5–)2–2.5 cm long, distinct, greenish light grey to brown on exposed portions, only the basal part exposed, the upper completely ensheathed by leaves. Leaves very spreading and recurving, 17–25(–35) per plant, flat to U-shaped, narrowly lanceolate to broadly linear-lanceolate, 27–32 cm long, 2–4.5(–6) cm wide, thick at the base, acute, spiral, slightly twisted especially the older leaves, smooth; spine 0.4–0.7 cm long, stout, horny, pale brown and more or less spotted; margins less than 0.5 mm wide, very narrow, pale brownish, with a greyish brown membrane that widens basally to almost 0.4(–0.5) cm wide. Inflorescence a terminal panicle, 55 cm high, ± 20–23 cm wide; peduncle 0.4–0.6 cm thick; branches 10–11(–15) cm long, basal branches with 3–5 secondary branchlets 4–6(–8) cm long; each branch and branchlet subtended by a bract 0.7–1.7 cm long, with a sharp spine; lower inflorescence bracts 1–2, lanceolate, 5.5–9 cm long, greenish grey but upper ones greyish brown (on drying), acute, with a sharp spine 0.5–0.6 cm long, greyish brown; flowers (2–)3(–4) per cluster, jointed at base; bracteoles 0.3 cm long, 0.2 cm wide, decreasing up the inflorescence; pedicel 0.2(–0.3) cm long. Flowers fugacious, whole perianth 1.2–1.4(–1.6) cm long, 0.1–0.2 cm wide; tube 0.4–0.6 cm long; lobes 0.5–0.6 cm long; style 1–1.3 cm long; filament 0.6–0.7 cm long, exserted, anthers 0.2 cm long; stigma globose. Fruit 1–3 per cluster, a globose orange berry, green when unripe.

KENYA. Kwale District: Voi–Mariakani near Taru, 10 Apr. 1973, *Pfennig* 1117!; Kilifi District: 36 km along Mombasa–Malindi road, 22 Mar. 1992, *Mbugua* 381! & 3 km from Mariakani on Nairobi road, 20 Mar. 1992, *Mbugua* 384!

TANZANIA. Handeni District: Kwa Mkono, 16 Feb. 1980, *Archbold* 2692!; Pangani District: Msubugwe Forest Reserve, 13 Mar. 1963, *Mgaza* 558!; Uzaramo District: near Kawe, 7 Mar. 1969, *Harris & Mwasumbi* 2835!

DISTR. **K** 7; **T** 3, 6; not known elsewhere

HAB. Evergreen coastal forest or evergreen bushland; may be locally common; 0–600 m

USES. No data

CONSERVATION NOTES. Least concern

SYN. *S. zanzibarica* Gérôme & Labroy in Bull Mus. Hist. Nat. Paris 9: 170, 173, fig. 19 (1903); N.E. Brown in K.B. 1915: 206 (1915); L.E. Newton in Ill. Handbook Succ. Plants 1: 272 (2001). Type: Tanzania, 15 Dec. 1909, *Sacleux* s.n. (K!, holo.), **syn. nov.**

NOTE. This is one of the species that has been mixed with *S. bagamoyensis* and even *S. powellii* for many years. This came about probablly because as N.E. Brown stated, he never saw a whole specimen of *S. bagamoyensis*, while it was very difficult to tell apart the specimens of *S. powellii*, especially when dry.

   S. arborescens with its spirally arranged, straight and rather high numbers of leaves per plant, with a transverse section of the leaf in an open U-shape, differs from *S. bagamoyensis* in that the latter has more leathery, flat light greenish (-yellowish) and generally larger leaves. At the juvenile stage the latter has distichous leaves which may become spiral towards the inflorescence (personal field observations). *S. bagamoyensis* also grows under thickets and forms more open stands than *S. arborescens* that grows even in the open and in very close stands. The distinction between all the above as far as inflorescence sizes are concerned is not yet quite clear - more material is needed for these to be studied in detail. *S. bagamoyensis* has leaves growing sideways and slightly backwards-curved while in *S. arborescens* they grow sideways and slightly upwards.

S. powellii N.E.Br. has more fibre content and the leaf is larger compared to *S. arborescens*. When *S. powellii* dries, the leaves shrink from side towards the centre forming grooves, while *S. arborescens* is narrower and the grooves are hardly noticeable. *S. powellii* is also very stiff and appears unbreakable while *S. arborescens* is flexible and usually more green, narrower and without a very pronounced green colour, narrowed and without very pronounced middle point of the leaf.

UNRESOLVED TAXA

Sansevieria subtilis *N.E.Br.* in K.B. 1915: 237, fig. 17 (1915); L.E. Newton in Ill. Handbook Succ. Plants 1: 271 (2001). Type: flowered at Kew in June 1913, from a specimen sent from Uganda, without locality, by *Dawe* s.n. (K, holo.)

Acaulescent, with creeping rootstock 8–11 mm in diameter. Leaves 2–4, erect or slightly recurving, linear-lanceolate, 55–70 cm long, 2.5–4.5 cm wide, 0.25 cm thick in mid-leaf, smooth, tapering at base into a 5–30 cm long channelled petiole, narrowing at apex to a 12–25 mm long soft subulate tip, green, lower surface sometimes with faint transverse bands, margin green. Inflorescence simple, spike-like, 38–53 cm long; flowers 2–3 per cluster; bracteoles lanceolate, 3–4.2 mm, acuminate; pedicel 4–7.5 mm. Flowers white, perianth tube 6–8.5 mm long, lobes recurved or slightly revolute, 10–13 mm long.

Collected from Uganda initially but grown at Kew from 1960-66 for cytological studies. Cyt: 63.804 2n=40. *Marchand, C.* 63.804 9.I.1964.; Basc West Ruwenzori, *Humbert* 3 1933.

INDEX TO DRACAENACEAE

New names validated in this part

Sansevieria aethiopica *Thunb.* subsp. **itumea** *Mbugua* subsp. nov.
Sansevieria gracilis *N.E.Br.* var. **humbertiana** (*Guill.*) *Mbugua*, stat. & comb. nov.
Sansevieria pfennigii *Mbugua* sp. nov.
Sansevieria subspicata *Baker* var. **concinna** (*N.E.Br.*) *Mbugua* stat. & comb. nov.

PLANTS PEOPLE
POSSIBILITIES

First published in 2007 by
Royal Botanic Gardens, Kew
Richmond, Surrey, TW9 3AB, UK
www.kew.org

ISBN 978 1 84246 187 7

British Library Cataloguing in Publication Data
A catalogue record for this book is available from the British Library

Design and typesetting by Margaret Newman,
Kew Publishing, Royal Botanic Gardens, Kew.

Printed in the UK by Hobbs the Printers

For information or to purchase all Kew titles please visit
www.kewbooks.com or email publishing@kew.org

All proceeds go to support Kew's work in saving the world's plants for life